Ben Stacy Jerrik (Ed.)

Marks Tey

Ben Stacy Jerrik (Ed.)

Marks Tey

West Mersea , Essex, Harlow , Epping, Matching Green

Part Press

Contents

Articles

Matching_Tye   1

Sawbridgeworth   3

Essex   7

Harlow   22

Epping   30

Matching_Green   36

Matching,_Essex   38

Rik_Mayall   40

References

Article Sources and Contributors   48

Image Sources, Licenses and Contributors   49

Matching_Tye

Matching Tye	
	Matching Tye
	Matching Tye shown within Essex
Population	635 [1]
Civil parish	Matching
District	Epping Forest
Shire county	Essex
Region	East
Country	England
Sovereign state	United Kingdom
Post town	HARLOW
Postcode district	CM17
Dialling code	01279
Police	Essex
Fire	Essex
Ambulance	East of England
EU Parliament	East of England
UK Parliament	Brentwood and Ongar

Matching Tye is a village which forms part of the civil parish of Matching, in the County of Essex, England. It is located 2.3 miles (3.7 km) East of Harlow, 2.9 Miles (4.8 km) South-East of Sawbridgeworth and 6.3 miles (10.4 km) North-East of Epping.[2]

Matching Parish make-up

- Matching
- Matching Green
- **Matching Tye**

Notable residents past and present

- Actor & Comedian, Rik Mayall[3] [4]

References

[1] Parish Profile : Matching (http://www.eppingforestdc.gov.uk/council_services/planning/census/matching.asp). Epping Forest District Council.
[2] Scalable map of the Matching Tye region (http://uk.epodunk.com/profiles/england/matching-tye/3001976.html)
[3] http://www.nndb.com/people/809/000060629/
[4] http://stason.org/TULARC/tv/rik-mayall/02-Personal-info-Rik-Mayall.html

External links

- Matching Parish Council (http://matchingcouncil.org.uk/index.php)

Sawbridgeworth

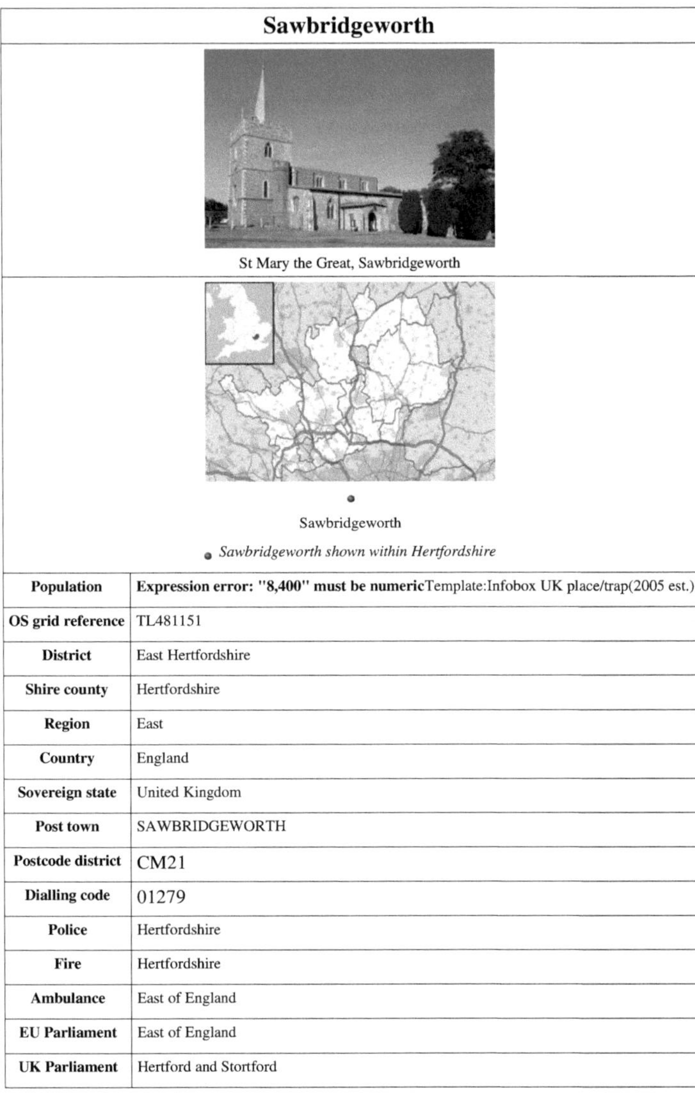

St Mary the Great, Sawbridgeworth

Sawbridgeworth

Sawbridgeworth shown within Hertfordshire

Population	Expression error: "8,400" must be numericTemplate:Infobox UK place/trap(2005 est.)
OS grid reference	TL481151
District	East Hertfordshire
Shire county	Hertfordshire
Region	East
Country	England
Sovereign state	United Kingdom
Post town	SAWBRIDGEWORTH
Postcode district	CM21
Dialling code	01279
Police	Hertfordshire
Fire	Hertfordshire
Ambulance	East of England
EU Parliament	East of England
UK Parliament	Hertford and Stortford

Sawbridgeworth is a small, mainly residential, town and also a civil parish in Hertfordshire, England.

Location

Sawbridgeworth is four miles south of Bishop's Stortford, twelve miles east of Hertford and nine miles north of Epping. It lies on the A1184 and has a railway station that links to Liverpool Street station in London. The river Stort flows through the east of the town, past the Maltings (now a small business and residential area).

Nearby towns and cities: Bishop's Stortford, Harlow

Nearby villages: High Wych, Spellbrook, Much Hadham

History

Prior to the Norman Conquest, most of the area was owned by the Saxon Angmar the Staller.

The Manor of "Sabrixteworde" (one of the many spellings previously associated with the town) was recorded in the Domesday Book. After the Battle of Hastings it was granted to Geoffrey de Mandeville by William the Conqueror. Local notables have included Sir John Leventhorpe, who was an executor of King Henry V's will, and Anne Boleyn, who was given the Pishiobury/Pishobury estate, located to the south of the town.

Much of the town centre is a conservation area; many of the buildings date from the Tudor, Stuart, and Georgian periods. Great St. Mary's church was originally built in the 13th century (although a church on the site existed in Saxon times) and includes a Tudor tower containing a clock bell (1664) and eight ringing bells, the oldest of which dates from 1749.[1] Historically, it is unclear where the apostrophe lies in "Great St. Mary's", and even which St. Mary it was, and why it is "Great". Ralph Jocelyn of Hyde Hall, who was twice Lord Mayor of London in the 15th century, is buried here; images of many of his family and other locals have been engraved on brass, and hence the church is popular for enthusiasts of brass rubbing.

a street entrance to Great Hyde Hall; it was known simply as Hyde Hall when it was held by the Jocelyn family in the 16th and 17th centuries

The town's prosperity came from the local maltings, owned by George Fawbert and John Barnard; in 1839 they set up the Fawbert and Barnard charity to fund local children and their education, funding a local infant school that still exists today.

Apart from the historic nature of the town, attractions include local river cruises in the summer, one annual fair held on Fair Green on the Sunday of the first May Bank Holiday and Carol singing on the green on Christmas Eve.

By the time of the Norman conquest, or soon after, Sawbridgeworth's rich farming land was fully developed for cultivation as was possible with the means available at the time: it was the richest village community in the country. It is, then, hardly surprising that many important medieval families had estates here. The land was divided amongst them, into a number of manors or distinct estates; the Lord of each manor had rights not only over this land but also over the people who farmed it. The number of manors increased during the Middle Ages, by a process of subinfeudation, that is the granting out of a part of an existing manor to a new owner so that the new manor was created. Many manors sprang from the original Domesday Book holding of the de Mandeville family. The first came to be called Sayesbury manor, from the de Say family who inherited it from the de Mandeville's in 1189. The many important people who held these manors built themselves houses with hunting parks around them; when they died their tombs enriched the parish church, so that today St. Mary's has one of the finest collections of church monuments in the country.

During World War II RAF Sawbridgeworth[2] operated Supermarine Spitfires, Westland Lysander, P-51 Mustang, and de Havilland Mosquitos.

Pronunciation

The name of the town is now almost universally pronounced in the obvious way, but this was not always the case. In the Middle Ages it is believed to have been pronounced "Sapserth", and since then the pronunciation has varied to include "Sapsa", "Sapster" and "Sapsworth", and even until the Second World War was pronounced "Sapsed". Current residents often use the casual abbreviated name "Sawbo".

Politics and local government

Sawbridgeworth is administered by East Herts district council. The local Town Council currently has 12 councillors, covering both Sawbridgeworth and Spellbrook.

Sawbridgeworth is twinned with:

* ▌▌ Bry-sur-Marne in France (1973)

In Parliament, it is in the Hertford and Stortford constituency. Since the election of May 2005 Sawbridgeworth is represented by Mark Prisk, a Conservative.

The village of Lower Sheering in Essex adjoins Sawbridgeworth, east of the railway station, along the Hertfordshire - Essex border. It shares the same postal code as Sawbridgeworth, although for local government purposes it comes under the Epping Forest district of Essex, with the Member of Parliament being Robert Halfon (Conservative).

Education

Sawbridgeworth has a secondary school, the Leventhorpe School, which also offers a public swimming pool and gym. There is also one primary school Mandeville, one junior school Reedings and one infant school Fawbert & Barnard in Sawbridgeworth.

Sport

Sawbridgeworth Town F.C., a non-league football club, plays at Crofters End.

Sawbridgeworth Cricket Club is one of the leading cricket clubs in the south of England, fielding five senior sides on a Saturday and seven colts sides, from ages nine to fifteen. The 1st XI plays in the Home Counties Premier Cricket League, and the other league sides play in the Hertfordshire Cricket League. The main ground is Town Fields, situated behind Bell Street. The second ground is at the Leventhorpe School.

Sawbridgeworth also has tennis and bowls clubs.

Local groups

Sawbridgeworth is home to 309 Squadron of the Air Training Corps.[3]

Transport

Sawbridgeworth railway station is situated on the London to Cambridge line. There is a ticket office and self service ticket machine, but there is no longer a permit to travel machine. There are also many bus services to the nearby towns of Harlow and Bishops Stortford. See Essex Bus Routes.

References

[1] Sawbridgeworth Church bells (http://www.sawbridgeworthchurch.com/bells.htm)

[2] http://merlinsroared.tripod.com/id3.html

[3] http://www.air-cadets-squadron-finder.org/air-cadets-squadron-details.php?sqn=0309--sawbridgeworth-air-training-corps-atc

External links

- Local council guide (http://www.eastherts.gov.uk/district/district_guide/dgsawbri.htm)
- Great St. Mary's church (http://www.sawbridgeworthchurch.com/)
- Leventhorpe School (http://www.leventhorpe.herts.sch.uk/)
- Sawbridgeworth Cricket Club (http://sawbridgeworthcc.hitscricket.com/home/)
- www.geograph.co.uk : Photos of Sawbridgeworth and surrounding area (http://www.geograph.org.uk/search.php?i=2680362)
- History of the Fire Brigade & photos of Sawbridgeworth and surrounding area (http://www.sawbridgeworthfirebrigade.co.uk:)

Essex

<table>
<tr><td colspan="2" align="center">Essex</td></tr>
<tr><td colspan="2" align="center">Motto of County Council: Essex Works.
For a better quality of life</td></tr>
<tr><td colspan="2" align="center">Geography</td></tr>
<tr><td>Status</td><td>Ceremonial and (smaller) non-metropolitan county</td></tr>
<tr><td>Origin</td><td>Historic</td></tr>
<tr><td>Region</td><td>East of England</td></tr>
<tr><td>Area
- Total
- Admin. council
- Admin. area</td><td>Ranked 11th
3670 km^2 (unknown operator: u'strong' sq mi)
Ranked 11th
3465 km^2 (unknown operator: u'strong' sq mi)</td></tr>
<tr><td>Admin HQ</td><td>Chelmsford</td></tr>
<tr><td>ISO 3166-2</td><td>GB-ESS</td></tr>
<tr><td>ONS code</td><td>22</td></tr>
<tr><td>NUTS 3</td><td>UKH33</td></tr>
<tr><td colspan="2" align="center">Demography</td></tr>
<tr><td>Population
- Total (2010 est.)
- Density
- Admin. council
- Admin. pop.</td><td>Ranked 6th
1,737,900
474 /km^2 (unknown operator: u'strong' /sq mi)
Ranked 2nd
1,412,900</td></tr>
<tr><td>Ethnicity</td><td>96.8% White
1.2% S. Asian</td></tr>
<tr><td colspan="2" align="center">Politics</td></tr>
</table>

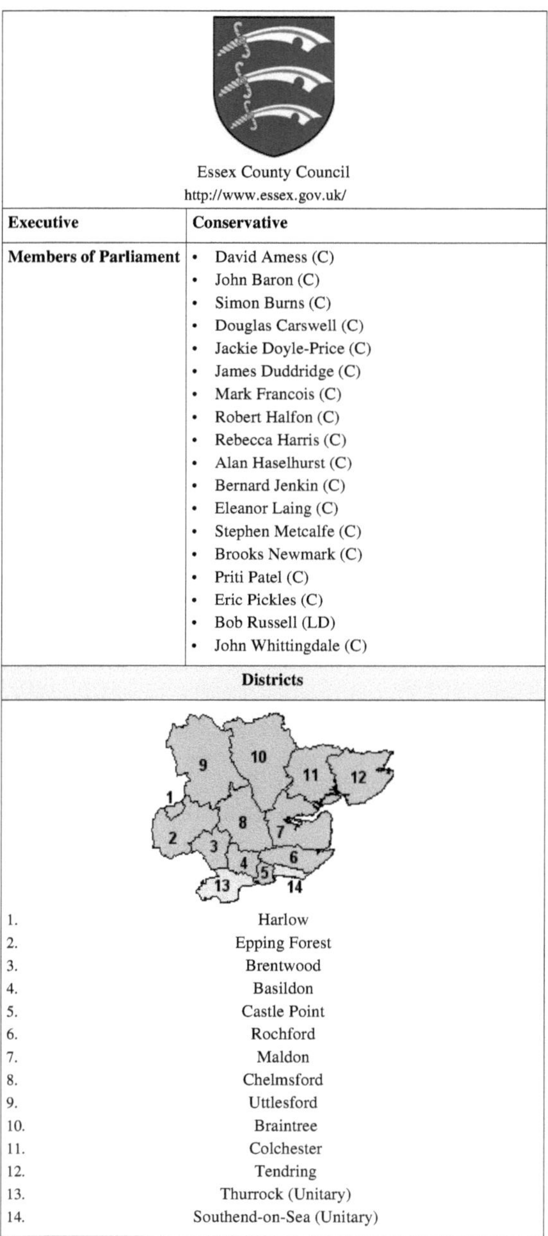

Essex County Council	
http://www.essex.gov.uk/	
Executive	**Conservative**
Members of Parliament	• David Amess (C) • John Baron (C) • Simon Burns (C) • Douglas Carswell (C) • Jackie Doyle-Price (C) • James Duddridge (C) • Mark Francois (C) • Robert Halfon (C) • Rebecca Harris (C) • Alan Haselhurst (C) • Bernard Jenkin (C) • Eleanor Laing (C) • Stephen Metcalfe (C) • Brooks Newmark (C) • Priti Patel (C) • Eric Pickles (C) • Bob Russell (LD) • John Whittingdale (C)
Districts	

1.	Harlow
2.	Epping Forest
3.	Brentwood
4.	Basildon
5.	Castle Point
6.	Rochford
7.	Maldon
8.	Chelmsford
9.	Uttlesford
10.	Braintree
11.	Colchester
12.	Tendring
13.	Thurrock (Unitary)
14.	Southend-on-Sea (Unitary)

Essex (🔊 /ˈɛsɪks/) is a ceremonial and non-metropolitan county in the East of England, and one of the home counties. It is located to the northeast of Greater London. It borders with Cambridgeshire and Suffolk to the north, Hertfordshire to the west, Kent to the South and London to the south west. Essex County Council is the principal local authority for much of the county, sharing functions with twelve district councils. The county town is

Chelmsford. The southern Essex boroughs of Thurrock and Southend-on-Sea are governed separately as unitary authorities. It was established in antiquity and formed the eastern portion of the Kingdom of Essex. Sections of the county closer to London are part of the Metropolitan Green Belt, which prohibits development. It is the location of the regionally significant Lakeside Shopping Centre and London Stansted Airport; and the new towns of Basildon and Harlow.

History

The name *Essex* originates in the Anglo-Saxon period of the Early Middle Ages and has its root in the Old English *Ēastseaxe* (i.e. the "East Saxons"), the eastern kingdom of the Saxons. Originally (in 527 AD) it occupied territory to the north of the River Thames, incorporating all of what later became Middlesex and most of what later became Hertfordshire, though its territory was later restricted to lands east of the River Lea.[1] Colchester in the north east of the county is Britain's oldest recorded town, dating back to before the Roman conquest, when it was known as Camulodunum and was sufficiently well-developed to have its own mint. Subsequently the Kingdom of Essex was subsumed into the Kingdom of England and Essex eventually became a county.

Essex County Council was formed in 1889. However the County Borough of West Ham and, from 1915, the County Borough of East Ham formed part of the county but were not under county council control.[2] A few Essex parishes were transferred to other counties. In earlier times, parts were transferred to Hertfordshire near Bishops Stortford and Sawbridgeworth. At the time of the main changes around the turn of the 19th/20th centuries; parts of Helions Bumpstead and Sturmer were transferred to Haverhill Suffolk, Kedington and Ballingdon-with-Brundon both went to Suffolk, and Great Chishill, Little Chishill and Heydon to Cambridgeshire. Southend-on-Sea also formed a county borough from 1914 to 1974.[3] Later still, part of Hadstock was transferred to Linton, Cambridgeshire, part of Ashdon to Bartlow, Cambridgeshire, part of Chrishall to Cambridgeshire, part of Great Horkesley to Nayland in Suffolk, and several other small territories were transferred to all of those three counties.

The boundary with Greater London was established in 1965 when East Ham and West Ham county boroughs and the Barking, Chingford, Dagenham, Hornchurch, Ilford, Leyton, Romford, Walthamstow and Wanstead and Woodford districts[2] were transferred to form the London boroughs of Barking, Havering, Newham, Redbridge and Waltham Forest. Essex became part of the East of England Government Office Region in 1994 and was statistically counted as part of that region from 1999, having previously been part of the South East England region. In 1998, the boroughs of Southend-on-Sea and Thurrock were separated from the shire county of Essex becoming unitary boroughs.

The county of Essex is divided into a number of local government districts. They are Harlow, Epping Forest, Brentwood, Basildon, Castle Point, Rochford, Maldon, Chelmsford, Uttlesford, Braintree, Colchester, Tendring, Thurrock, and Southend-on-Sea.[4] The last two boroughs are unitary authorities which form part of the county but are not administered by the county council.[5] Essex Police also covers the two unitary authorities.[6]

The county council meets at the County Hall, in Chelmsford. Before 1938, it regularly met in London near Moorgate, which had been more convenient than any place in the county. It currently has 75 elected councillors. Before 1965, the number of councillors reached over 100. The County Hall, which dates largely from the mid-1930s, and is decorated with fine artworks of that period, mostly the gift of the family who owned the textile firm, Courtaulds, was made a listed building in 2007.

Geography

The highest point of the county of Essex is Chrishall Common near the village of Langley, close to the Hertfordshire border, which reaches 482 feet (**unknown operator: u'strong'** m). The ceremonial county of Essex is bounded to the south by the River Thames and its estuary (a boundary shared with Kent); to the southwest by Greater London; to the west by Hertfordshire with the boundary largely defined by the River Lea and the Stort; to the northwest by Cambridgeshire; to the north by Suffolk, a boundary mainly defined by the River Stour; and to the east by the North Sea.

The plaza of the new town of Basildon

The pattern of settlement in the county is diverse. The Metropolitan Green Belt has effectively prevented the further sprawl of London into the county, although it contains the new towns of Basildon and Harlow, originally developed to resettle Londoners following the destruction of London housing in World War II but since much expanded. Epping Forest also acts as a protected barrier to the further spread of London. Because of its proximity to London and the economic magnetism which that city exerts, many of Essex's settlements, particularly those on or within driving distance of railway stations, function as dormitory towns or villages where London workers raise their families.

Part of the south east of the county, already containing the major population centres of Basildon, Southend and Thurrock, is within the Thames Gateway and designated for further development. Parts of the south west of the county such as Buckhurst Hill and Chigwell are contiguous with Greater London and are included in the Greater London Urban Area. A small part of the south west of the county (Sewardstone), is the only settlement outside Greater London to be covered by a postcode district of the London post town (E4). To the north of the green belt, with the exception of major towns such as Colchester and Chelmsford, the county is rural, with many small towns, villages and hamlets largely built in the traditional materials of timber and brick, with clay tile or thatched roofs.

The village of Finchingfield in north Essex

Economy

Industry is largely limited to the south of the county, with the majority of the land elsewhere being given over to agriculture. Harlow is a centre for electronics, science and pharmaceutical companies, while Chelmsford is the home of Marconi (now called telent plc and owned by Ericsson of Sweden since 2005), Basildon home to New Holland Agriculture's European headquarters and Brentwood home to the Ford

Skyline of Southend-on-Sea

Motor Company's European HQ. Loughton is home to a production facility for British and foreign banknotes. Chelmsford has been an important location for electronics companies since the industry was born, is also the location for a number of insurance and financial services organisations, and is the home of the soft drinks producer Britvic. Other businesses in the county are dominated by light engineering and the service sector. Colchester is a garrison town, and the local economy is helped by the Army's personnel living there. Basildon is the location of State Street Corporation's United Kingdom HQ International Financial Data Services, and remains heavily dependent on London for employment, due to its close proximity and direct transport routes. Southend-on-Sea is home to the Adventure Island theme park and is one of the few still growing British Seaside resorts, benefiting from direct, modern rail links from Fenchurch Street railway station and Liverpool Street station (placing housing in high demand, especially for financial services commuters), which thereby maintains the town's commercial and general economy.

Parts of Eastern Essex suffer from high levels of deprivation, with one of the most highly deprived wards being in Clacton.[7] In the Indices of deprivation 2007, Jaywick was identified as the most deprived Lower Super Output Area in Southern England.[8] Unemployment was estimated at 44% and many homes were found to severely lack basic amenities. The Brooklands and Grasslands area of Jaywick were found to be the third most deprived area in England; only Liverpool and Manchester rated higher. In contrast, however, South West Essex is a mostly affluent part of Eastern England, forming part of the London commuter belt. South West Essex has a large middle class and the area is widely known for its Independent schools. In 2008, *The Daily Telegraph* found Brentwood and Ingatestone to be the 19th and 14th richest towns in the UK respectively.[9]

Transport

The main airport in Essex is London Stansted Airport, serving destinations in Europe and Asia. The Lib-Con coalition government formed in May 2010 has agreed to resist an additional runway at Stansted, so curtailing the operator's ambitions for expansion. London Southend Airport, once one of Britain's busiest airports, opened a new runway extension, terminal building and railway station in March 2012.[10] The station is on the Shenfield to Southend Line, with a direct link to the capital. Currently the airport offers scheduled flights to Ireland, the Channel Islands and multiple destinations in Europe. There are several smaller airfields, some of which owe their origins to

London Stansted Airport, in the north west of the county

military bases built during World War I or World War II. These are popular for pleasure flights or flying lessons; examples include Clacton Airfield, Earls Colne Airfield, and Stapleford Aerodrome.

The Port of Tilbury is one of Britain's three major ports, while the port of Harwich links the county to the Hook of Holland and Esbjerg (a service to Cuxhaven Having discontinued in December 2005). Plans have been approved to build the UK's largest container terminal at Shell Haven in Thurrock and although opposed by the local authority and environmental and wildlife organisations, it now seems increasingly likely to be developed.[11] [12] [13]

Despite the existence of the Dartford Road Crossing to Dartford, Kent, across the Thames River, a ferry for pedestrians to Gravesend, Kent, still operates from Tilbury during limited daily hours, and ferries for pedestrians operate across some of Essex's rivers and estuaries during spring and summer. The M25 and M11 motorway both cross the county in the extreme south and west, linking those parts of the county with Kent, Hertfordshire and Cambridge. The A127 and A13 trunk roads are important radial routes connecting London and the M25 to the south of Essex. The A12 runs across the county from the south west to the north east and not only carries traffic within Essex but also traffic between London and Suffolk, east Norfolk and the ports of Felixstowe and Harwich.

Queen Elizabeth II Bridge spanning the Thames from West Thurrock, Essex, to Dartford, Kent

There is an extensive public transport network. The main railway routes in Essex include two lines from the City of London to Southend-on-Sea, operated by c2c from Fenchurch Street railway station (including a route via Tilbury) and by Greater Anglia from Liverpool Street station; the Great Eastern Main Line from Liverpool Street connecting to Harwich and onwards into Suffolk and Norfolk; and the West Anglia Main Line from Liverpool Street linking to Stansted and onwards into Cambridgeshire. The Epping Forest district is served by the London Underground Central Line. The routes operated by Greater Anglia were formerly operated by National Express East Anglia and have been previously branded as 'one'. There are also a number of branch lines including the Sunshine Coast Line linking Colchester to the seaside resorts of Clacton-on-Sea and Walton-on-the-Naze via the picturesque towns of Wivenhoe and Great Bentley. The Crouch Valley Line, another branch line, links Wickford to a number of riverside communities via South Woodham Ferrers and Burnham-on-Crouch to Southminster.

South Essex Rapid Transit is a proposed public transport scheme which would provide a fast, reliable public transport service in, and between, Thurrock, Basildon and Southend.[14]

Education

Further information: List of schools in Essex, List of schools in Southend-on-Sea, List of schools in Thurrock, and List of primary schools in Essex

Education in Essex is substantially provided by three authorities being Essex County Council and the two unitary authorities, Southend-on-Sea and Thurrock. In all there are some 90 state secondary schools provided by these authorities, the majority of which are comprehensive, although one in Uttlesford, two in Chelmsford, two in Colchester and four in Southend-on-Sea are selective. There are also various Independent Schools providing education in Essex.[15] [16]

The University of Essex, which was established in 1963, is located just outside Colchester, with two further campuses in Loughton and Southend-on-Sea. University Campus Suffolk, with a main campus in Ipswich and five centres in the counties of Norfolk and Suffolk, is a joint venture between University of Essex and University of East Anglia. Anglia Ruskin University was awarded university status in 1992 and has campuses in Chelmsford and Cambridge.

Culture

The County's coat of arms comprises three Saxon seax knives (although looking rather more like scimitars) arranged on a red background; the three-seax device is also used as the official logo of Essex County Council having been granted as such in 1932.[17] The emblem was attributed to Anglo-Saxon Essex in Early Modern historiography. The earliest reference the arms of the East Saxon kings was by Richard Verstegan, the author of *A Restitution of Decayed Intelligence* (Antwerp, 1605), claiming that "Erkenwyne king of the East-Saxons did beare for his armes, three [seaxes] argent, in a field gules". There is no earlier evidence substantiating Verstegan's claim, which is an anachronism for the Anglo-Saxon period seeing that heraldry only evolved in the 12th century, well after the Norman conquest. John Speed in his *Historie of Great Britaine* (1611) follows Verstegan in his descriptions of the arms of Erkenwyne, but he qualifies the statement by adding "as some or our heralds have emblazed".[17]

Depiction of the first king of the East Saxons, Æscwine, his shield showing the three seaxes emblem attributed to him (from John Speed's 1611 *Saxon Heptarchy*).

Essex is also home to the Dunmow Flitch Trials, a traditional ceremony that takes place every four years and consists of a test of a married couple's devotion to one another. A common claim of the origin of the Dunmow Flitch dates back to 1104 and the Augustinian Priory of Little Dunmow, founded by Lady Juga Baynard. Lord of the Manor Reginald Fitzwalter and his wife dressed themselves as humble folk and begged blessing of the Prior a year and a day after marriage. The Prior, impressed by their devotion bestowed upon them a Flitch of Bacon. Upon revealing his true identity, Fitzwalter gave his land to the Priory on the condition a Flitch should be awarded to any couple who could claim they were similarly devoted. By the 14th century, the

The Hay Wain by John Constable shows the Essex landscape on the right bank

Dunmow Flitch Trials appear to have achieved a significant reputation outside the local area. The author William Langland, who lived on the Welsh borders, mentions it in his 1362 book *The Vision of Piers Plowman* in a manner that implies general knowledge of the custom among his readers.[18]

Landmarks

Over 14,000 buildings have listed status in the county, and around 1000 of those are recognised as of Grade I or II* importance.[19] The buildings range from the 7th century Saxon church of St Peter-on-the-Wall, to the Royal Corinthian Yacht Club which was the United Kingdom's entry in the "International Exhibition of Modern Architecture" held at the Museum of Modern Art in New York City in 1932.

The church of St Peter-on-the-Wall, Bradwell-on-Sea.

The Grade I listed Hedingham Castle.

Thaxted Guildhall dating from around 1450.

The 17th century Audley End House in Saffron Walden.

The Royal Corinthian Yacht Club, Burnham-on-Crouch.

Colchester Castle

Hylands House

Southend Pier

Places of interest

Key	
✝	Abbey/Priory/Cathedral
AL	Accessible open space
⛩	Amusement/Theme Park
🏰	Castle
👤	Country Park
⊞	English Heritage
🌲	Forestry Commission

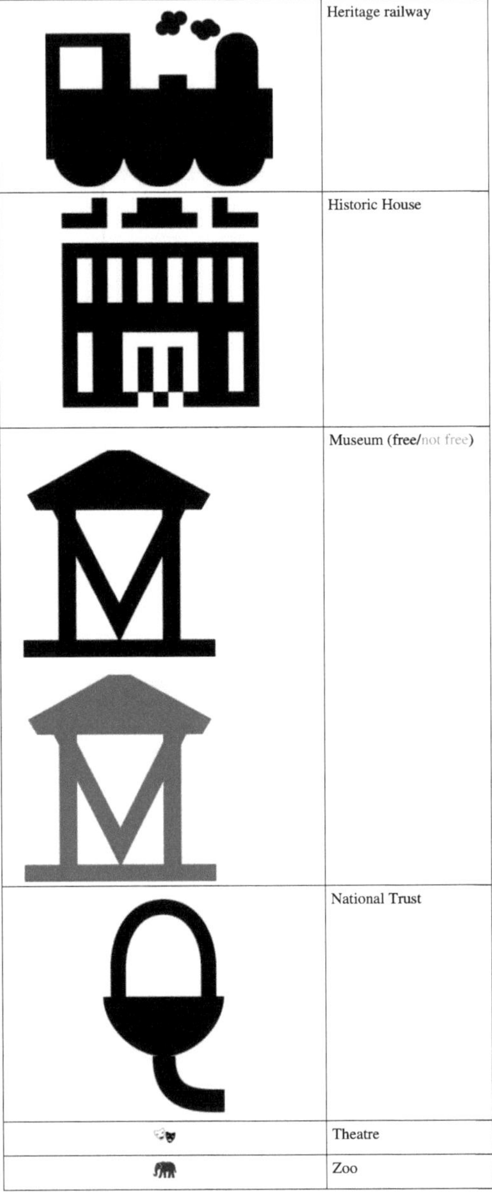

	Heritage railway
	Historic House
	Museum (**free**/not free)
	National Trust
	Theatre
	Zoo

- Abberton Reservoir
- Ashingdon (The site of the Battle of Ashingdon in 1016)
- Audley End
- Clacton-On-Sea

- Colchester Castle [20]
- Chelmsford Cathedral ✝
- Colchester Zoo 🐘
- Colne Valley Railway

- East Anglian Railway Museum 🏛
- Epping Forest

- Frinton-on-Sea

- Great Bentley   Home to the
  Largest Village green in England
- Harlow New Town

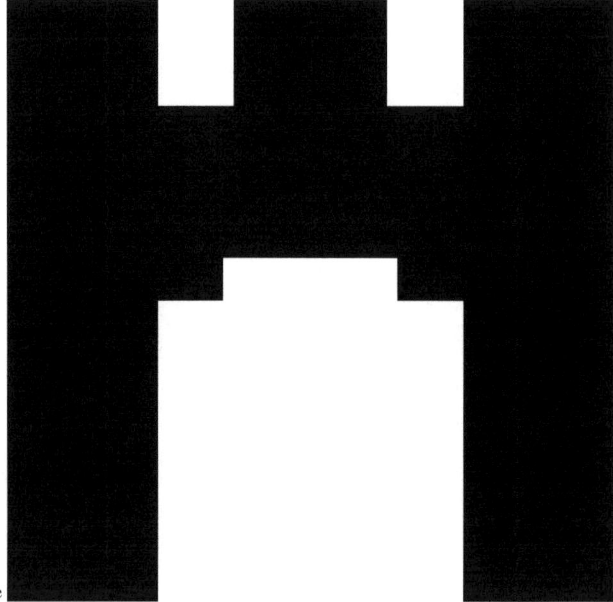

- Hedingham Castle
- Ingatestone Hall

- Kelvedon Hatch (Secret Nuclear Bunker)

- Loughton
- Maldon Historic market town site of the Battle of Maldon
- Mangapps Railway Museum

(Burnham-on-Crouch)
- Marsh Farm Country Park (South Woodham Ferrers)
- Mersea Island
- Mistley Towers

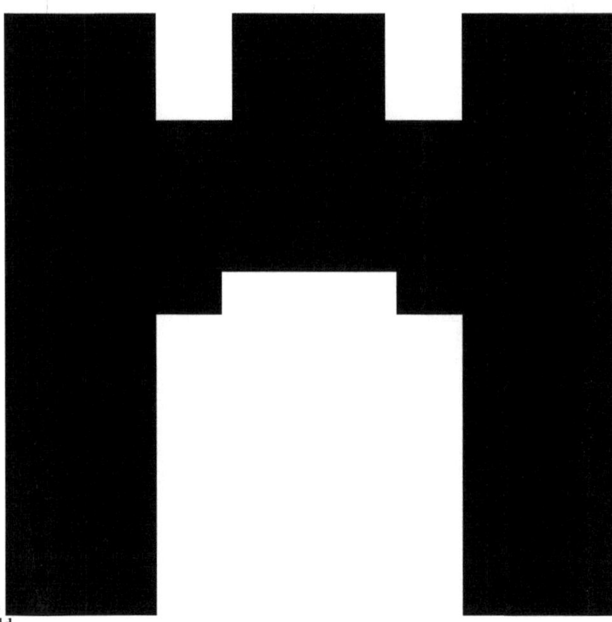

- Mountfitchet Castle
- North Weald Airfield
- Orsett Hall
- River Thames **AL**
- St Peter-on-the-Wall ✝

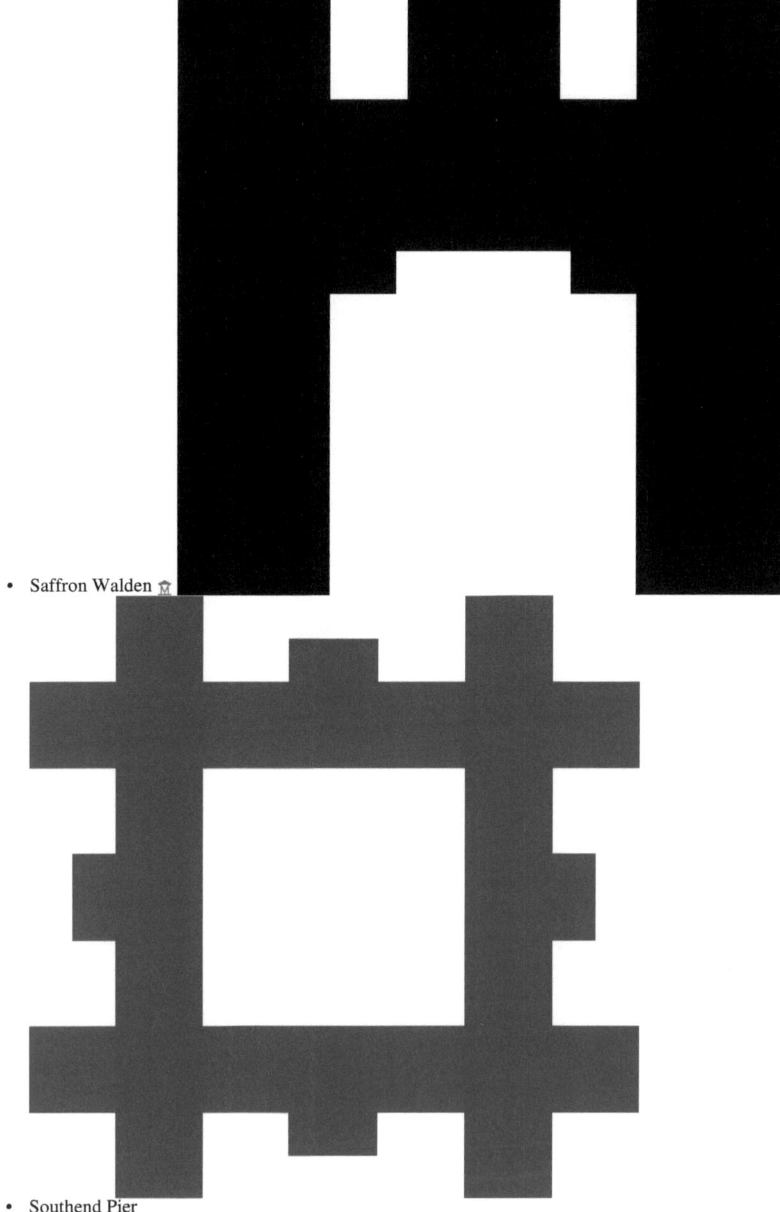

- Saffron Walden 🏛
- Southend Pier
- Thaxted
- University of Essex (Wivenhoe Park, Colchester)
- Waltham Abbey ✝

Notable persons

- *See Category:People from Essex.*

Sister counties and regions

- Jiangsu, China
- Picardy, France
- Thuringia, Germany
- Henrico County, Virginia
- Accra, Ghana

See also

- The Earl of Essex
- List of Lord Lieutenants of Essex
- List of High Sheriffs of Essex
- Custos Rotulorum of Essex - Keepers of the Rolls
- Historical list of MPs of Essex constituency
- Q Camp: WWII camp in Essex
- List of civil parishes in England

Notes and references

[1] Vision of Britain (http://www.visionofbritain.org.uk/bound_map_page.jsp?first=true&u_id=10001079&c_id=10001043) – Essex ancient county boundaries map

[2] Vision of Britain (http://www.visionofbritain.org.uk/relationships.jsp?u_id=10186092&c_id=10001043) – Essex admin county (historic map (http://www.visionofbritain.org.uk/bound_map_page.jsp?first=true&u_id=10186092&c_id=10001043))

[3] Vision of Britain (http://www.visionofbritain.org.uk/relationships.jsp?u_id=10135618) – Southend-on-Sea MB/CB

[4] Essex County Council (http://www.essexcc.gov.uk/vip8/ecc/ECCWebsite/dis/guc.jsp?channelOid=71101&guideOid=93686&guideContentOid=93690) – District or Borough Councils

[5] OPSI (http://www.opsi.gov.uk/si/si1996/Uksi_19961875_en_1.htm) – The Essex (Boroughs of Colchester, Southend-on-Sea and Thurrock and District of Tendring) (Structural, Boundary and Electoral Changes) Order 1996

[6] OPSI (http://www.opsi.gov.uk/si/si1997/19971847.htm) – The Essex (Police Area and Authority) Order 1997

[7] http://www.chelmsford.anglican.org/did-you-know-deprivation-in-chelmsford-diocese.html

[8] http://www.clactonandfrintongazette.co.uk/news/colchester/1901353.Jaywick__Village____third_most_deprived_area_in_UK___/

[9] "Britain's richest towns: 20 – 11" (http://www.telegraph.co.uk/property/3361038/Britains-richest-towns-20-11.html). *The Daily Telegraph* (London). 18 April 2008. .

[10] Topham, Gwyn (5 March 2012). "London Southend airport: flying under the radar (and to the left of the pier)" (http://www.guardian.co.uk/world/2012/mar/04/london-southend-airport?INTCMP=SRCH). London: The Guardian. . Retrieved 5 March 2012.

[11] Portswatch: Current Port Proposals: London Gateway (Shell Haven) (http://www.foe.co.uk/campaigns/transport/portswatch/port_proposals/london_gateway.html). Retrieved 15 April 2009.

[12] Thurrock Council. (2003-02-26). Shell Haven public inquiry opens (http://www.thurrock.gov.uk/news/content.php?page=story&ID=134). Retrieved 15 April 2009.

[13] Dredging News Online. (2008-05-18). Harbour Development, Shell Haven, UK (http://www.sandandgravel.com/news/article.asp?v1=10983). Retrieved 15 April 2009.

[14] "FAQ" (http://www.sert.org.uk/faqs.asp). .

[15] Essex County Council. (2006). Secondary School Information (http://www.essexcc.gov.uk/vip8/ecc//setSchoolType.do?stype=Secondary). Retrieved 15 April 2009.

[16] Independent Schools Directory. (2009). Independent Schools in Essex (http://www.independentschools.com/england/independent-schools-in-essex.html). Retrieved 15 April 2009.

[17] Robert Young. (2009). Civic Heraldry of England and Wales. Essex (http://www.civicheraldry.co.uk/essex.html). Retrieved 16 April 2009.

[18] Dunmow Flitch Trials website. Retrieved 12 April 2010. (http://www.dunmowflitchtrials.co.uk/history/)

[19] Bettley, James. (2008). Essex Explored: Essex Architecture. (http://www.realessex.co.uk/discover/historic/EssexArchitecture.aspx) Essex County Council. Retrieved 15 April 2009.

[20] "Colchester Castle Museum-Index" (http://www.colchestermuseums.org.uk/castle/castle_index.html). Colchestermuseums.org.uk. . Retrieved 23 April 2010.

External links

- Essex (http://www.dmoz.org/Regional/Europe/United_Kingdom/England/Essex/) at the Open Directory Project
- Visit Essex (http://www.visitessex.com/)
- Essex County Council (http://www.essexcc.gov.uk/)
- Seax – Essex Archives Online (http://seax.essexcc.gov.uk/)

Harlow

<table>
<tr><td colspan="2" align="center">Harlow</td></tr>
<tr><td colspan="2">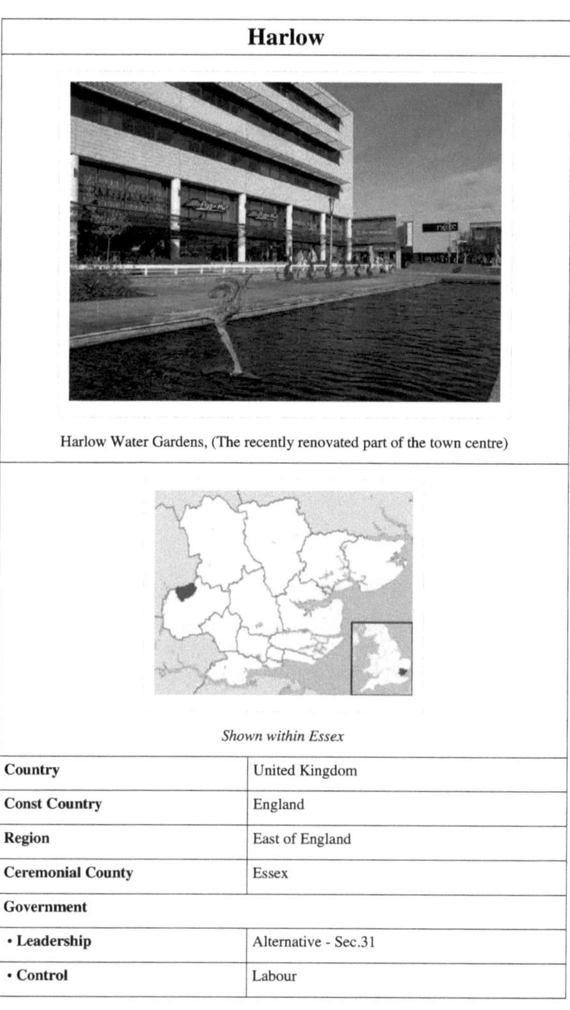
Harlow Water Gardens, (The recently renovated part of the town centre)

Shown within Essex</td></tr>
<tr><td>Country</td><td>United Kingdom</td></tr>
<tr><td>Const Country</td><td>England</td></tr>
<tr><td>Region</td><td>East of England</td></tr>
<tr><td>Ceremonial County</td><td>Essex</td></tr>
<tr><td>Government</td><td></td></tr>
<tr><td>• Leadership</td><td>Alternative - Sec.31</td></tr>
<tr><td>• Control</td><td>Labour</td></tr>
</table>

• **MP**	Robert Halfon
Area	
• District	**unknown operator: u'strong'** sq mi (30.54 km^2)
Population	
• District	Ranked 285th 81,700
• Density	**unknown operator: u'strong'**/sq mi (2675/km^2)
• Ethnicity[1]	92.5% White 2.5% Asian 1.8% Black 1.6% Chinese or Other 1.5% Mixed
ONS code	22UJ
Website	http://www.harlow.gov.uk/

Harlow is a new town and local government district in Essex, England. It is located in the west of the county and on the border with Hertfordshire, on the Stort Valley, The town is near the M11 motorway and forms part of the London commuter belt.The district has a current population of 78,889 (2010 estimate).[2]

History

Etymology

There is some dispute as to where the placename Harlow derives from. One theory is that it derives from the Anglo-Saxon words 'here' and 'hlaw', meaning "army hill", probably to be identified with Mulberry Hill, which was used as the moot or meeting place for the district.

The other theory is that it derives from the words 'here' and 'hearg', meaning "temple hill/mound", probably to be identified with an Iron Age burial mound, later a Roman temple site on River Way.

The original village, mentioned in the Norman Domesday Book, developed as a typical rural community around what is now known as Old Harlow, with many of its buildings still standing.

Early history

The earliest deposits are of a Mesolithic (circa 10,000 BC) hunting camp excavated by Davey in Northbrooks in the 1970s(Unpublished) closely followed by the large and unexcavated deposits of Neolithic flint located at Gilden Way. These deposits are mostly known of due to the large numbers of surface bound worked flint ; indeed there is substanital amounts to speculate on organised working of flint in the area. Large amounts of debetage litter the area and tools found include Axeheads, hammers, blades, dowles and other boring tools and multipuprpose flints such as scrapers. An organised Field Walk in the late 1990s by Bartlett(Unpublished)indicates that most of the area, some 80 hectares, produces worked flint from the Neolithic to the Bronze Age with a smattering of Mesolithic ; so basically it indicates organised industry from 5000 BC to 2000 BC. Indeed the deposits are so large and dispersed that any major archaeological work in the area will have to take this into consideration before any ground work is started.

The New Town

The new town was built after World War II to ease overcrowding in London at the same time as the similar orbital developments of Basildon, Stevenage, and Hemel Hempstead. The master plan for the new town was drawn up in 1947 by Sir Frederick Gibberd.[3] [4] The development incorporated the market town of Harlow, now a neighbourhood known as Old Harlow, and the villages of Great Parndon, Latton, Tye Green, Potter Street, Churchgate Street, Little Parndon, and Netteswell. The town is divided into neighbourhoods, each self supporting with their own shopping precincts, community facilities and pub. Gibberd invited many of the country's leading post-war architects to design buildings in the town, including Philip Powell and Hidalgo Moya, Leonard Manasseh, Michael Neylan, E C P Monson, Gerard Goalen, Maxwell Fry, Jane Drew, Graham Dawbarn and William Crabtree. Harlow has one of the most extensive cycle track networks in the country, connecting all areas of the town to the town centre and industrial areas. The cycle network is composed mostly of the original pre-new town roads.

The town is notable being the location of Britain's first pedestrian precinct,[5] and first modern-style residential tower block, The Lawn,[6] [7] constructed in 1951; it is now a Grade II listed building. Gibberd's tromp-l'oeil terrace in Orchard Croft and Dawbarn's maisonette blocks at Pennymead are also notable, as is Michael Neylan's pioneering development at Bishopsfield. The first neighbourhood, Mark Hall, is a conservation area. From 1894 to 1955, the Harlow parish formed part of the Epping Rural District of Essex.[8] From 1955 to 1974, Harlow was an urban district.[9]

The town centre, and many of its neighbourhood shopping facilities have undergone major redevelopment, along with many of the town's original buildings. Subsequently, many of the original town buildings, including most of its health centres, the Staple Tye shopping centre, and many industrial units have been rebuilt. GIbberd's original town hall, a landmark in the town, has been demolished and replaced by a new civic centre and shopping area.

Redevelopment

The town has already experienced expansion. The first of which was the "mini expansion" that was created by the building of the Sumners and Katherines estates in the mid to late seventies to the west of the existing town. Since then Harlow has further expanded with the Church Langley estate completed in 2005, and its newest neighbourhood Newhall nearing completion. The Harlow Gateway Scheme is currently underway, with the relocation of the Harlow Football Stadium to Barrow's Farm in early 2006, and the building of a new hotel, apartments, and a restaurant adjacent to the railway station being complete. The next stage of this scheme involves the completion of the 530 eco-homes being built on the former sports centre site, and the centre's relocation to the nearby former college playing field site.

Other major developments under consideration include both a northern and southern bypass of the town, and significant expansion to the north, following the completed expansion to the east. The Harlow North[10] plans, currently awaiting permission, involve an extension of the town across the floodplains on the town's northern border, into neighbouring Hertfordshire. The plan was supported by former MP Bill Rammell, all three political groups on Harlow Council, and the East of England Regional Assembly. It is opposed by Hertfordshire Council Council, East Herts Council, Mark Prisk, MP for Hertford, and Stortford in whose constituency the development would be and all the parishes concerned. The opposition is coordinated by a local group based in neighbouring East Hertfordshire.[11] An attempt to have Harlow North designated an "Eco Town" was rejected by the Minister for Housing, Caroline Flint, MP in April 2008

The south of the town centre also underwent major regeneration, with the new civic centre being built and the town's famous Water Gardens being redeveloped, a landscape listed by English Heritage. Their intended effect is somewhat spoiled by the abutment of a range of new shops, a major superstore, and several restaurants and cafés. It is likely that this development will be continued throughout the rest of the shopping district, with plans awaiting planning permission to be granted.

Economy

Harlow was originally expected to provide a majority of employment opportunities in manufacturing, with two major developments of The Pinnacles and Templefields providing the biggest employers in the region; as with the rest of the country, this manufacturing base has declined and Harlow has had to adjust.

The original manufacturing took the form of a biscuit factory, on the Pinnacles. Owned and run as a Co-Op, it provided employment to the town for over 50 years, before closing in 2002. It has since been demolished and the site is now small industrial units. At its peak, the factory employed over 500 people. At the time of its closure, the owner was Burton's Foods Ltd. An £8million production line – installed in 1999 – was left to rust in the car park upon the closure of the factory.

Raytheon and GlaxoSmithKline both have large premises within the town. Nortel also had a large site on the eastern edge of the town, acquired when STC was bought in 1991, and it was here that Charles K. Kao developed optical fibre data transmission. Nortel still has a presence, but it is much reduced. One of Europe's leading online golf stores, Onlinegolf, is also based in Harlow.

Unemployment is frequently around 10%, higher than the national average in the UK. Harlow also has a large number of people in social housing, almost 30%[2] of dwellings being housing association and local authority owned, and many more privately rented.

Politics

See also: Harlow (UK Parliament constituency), Harlow local elections

Labour MP Bill Rammell was reelected in the 2005 general election, with a majority of only 97 after considerable gains by the Conservative Party since the 1997 and 2001 elections but lost to Robert Halfon, Conservative, in the 2010 general election. Prior to the 2008 Council elections, no party had overall control of the local authority, which was run by a coalition of Liberal Democrats and Labour Party councillors. However, since the elections, the authority is under Conservative control.

Transport

Rail

Harlow is served by two railway stations, Harlow Town railway station and Harlow Mill railway station. There is also a bus service to Epping tube station on the London Underground.

Road

Harlow is reached from junction 7 of the M11 motorway, which runs from London to Cambridge, placing it within a short distance of Stansted Airport and the A120 and the orbital M25 motorway. Running through the town is the A414, a major road from Hertford to Chelmsford and linking the town with the A10 to the west. This road is often a cause of major congestion to the town and is

A WAGN EMU travels through Harlow station in 2001.

awaiting a decision of both a southern and northern bypass to the town, with the Harlow North proposal including the latter as part of its bid to secure planning permission for 8,000 homes to the North of the town. It is unlikely to be built in the near future however. Another major road running from Harlow is the A1184, which also leads to the nearby town of Bishop's Stortford.

Air

Bishop's Stortford is the closest large town to London Stansted Airport, though Harlow is only 10 miles from this major transport hub, and therefore provides several hundred airport employees. BAA , the airport operator, withdrew a planning application for a second runway after the General Election of 2010, when all major political parties opposed it. .

Bus

Harlow has an extensive bus network and serves as a regional hub for the local area, with operators such as Arriva East Herts & Essex, SM Coaches, Roadrunner Coaches, Centrebus, and TWH Bus & Coach.

Future transport plans

Harlow bus station

Harlow First Avenue Multi-Modal Corridor

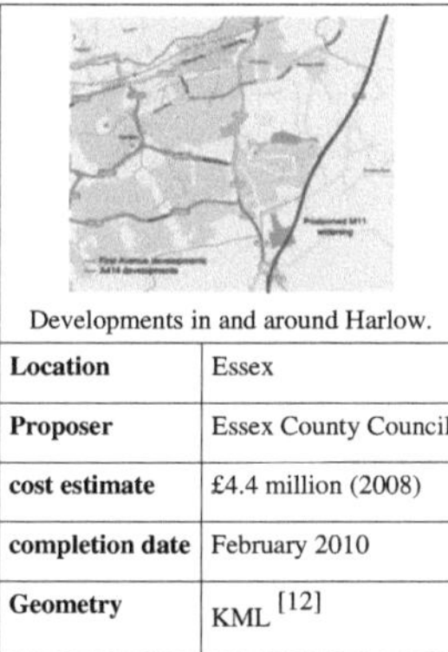

Developments in and around Harlow.	
Location	Essex
Proposer	Essex County Council
cost estimate	£4.4 million (2008)
completion date	February 2010
Geometry	KML [12]

Essex County Council is involved in development to Harlow's First Avenue, which is intended to reduce congestion and create better transport connections between the Newhall housing developments. The scheme was implemented in two phases, each phase focusing on developing First Avenue on either side of Howard Way. Phase two has an estimated cost of £4.4 million and is due to be completed in early 2010, phase one is already complete and is listed as having had £3.6 million of funding from the Community Infrastructure Fund (CIF).[13] The scheme includes construction of a shared use cycleway and development to the bus service along first avenue and into the Newhall development site where 'high quality bus' services between Harlow town centre and Harlow Town Railway station are listed as part of the development.

Healthcare

Harlow is served by the NHS Princess Alexandra Hospital, situated on the edge of The High, which is the main Town Centre area of Harlow. This hospital has a 24 hour Accident & Emergency and Urgent Care Centre.

There is also a private hospital called The Rivers, which is located on the outskirts of Harlow. It is run by the Capio group alongside the Jacobs Centre which serves neurological patients.

Education

Harlow contains six secondary schools, most of which now have specialist status, and one College.

- Mark Hall Specialist Sports College - Sports College[14]
- St Mark's Catholic School - Business & Enterprise Specialist[15] (Also has a sixth form as part of the school)
- Burnt Mill Academy - Performing arts College[16]
- Stewards Academy - Science Specialist[17]
- Passmores Academy - Technology College[18]
- Harlow College - College[19]
- Saint Nicholas School

Brays Grove School closed down in 2008 due to falling numbers of school aged students in the town. Passmores School and Technology College moved into a brand new £23 million school in September 2011 on the site of the old Brays Grove School.[20]

In the 1980s a further two secondary schools were closed, Latton Bush (now a commercial centre and recreational centre) and Netteswell (now forms part of the Harlow College Campus)[21] is a major further educational centre, covering GCSE's, A-Levels, and many vocational subjects including Hair & Beauty Therapy, Construction, Mechanics, ICT, and a new centre for Plumbing due to open. The college is currently under major regeneration and is due to open a new university centre in partnership with Anglia Ruskin University, covering mostly Foundation degrees in a variety of subjects relevant to local employers needs.

Memorial University of Newfoundland also has a small international campus located in Old Harlow.

Sport and leisure

Harlow Rugby Football Club play their home games at Ram Gorse in the town. The first team plays in the London & South East Division II North East league.

Harlow has 4 cricket clubs.

Harlow Town Cricket Club was formed in 1960 as Stort Cricket Club and used the new Sportcentre ground but now plays at Ash Tree Field.[22] The club now competes in Shepherad neame League and runs 5 league, 7 colts sides and 3 veterans teams making it the biggest cricket club in Harlow in terms of size.http://stort.play-cricket.com/content/view.asp?id=10173577&cid=202</ref>

Harlow Cricket Club traces its history back to 1774.[22] The club plays league cricket in the Essex Shepherd Neame League from its Old Harlow ground of Marigolds. Near neighbour Potter Street & Church Langley Cricket Club play in the Herts & Essex League. Netteswell & Burnt Mill Cricket Club dates back to 1889 and plays friendly cricket against local clubs.

The town's football team Harlow Town F.C. play in the Ryman Division One North. In October 2006 they moved into their new stadium at Barrows Farm, and their old ground at the Harlow Sportcentre has been demolished to make way for new housing facilities as part of the Gateway Scheme, which will also see a brand new sports centre complex built in the centre of the town, on the former Harlow College playing field.

The town was the site of the UK's first purpose-built sports centre, Harlow Sports Centre, in 1960. The building is due to be replaced on 1 July 2010 [23] by the state-of-the-art Harlow Leisure Park, built near Harlow College as part of the Gateway Project. Harlows 'Leisurezone' opened on 23 June 2010, with new dry and wet sports facilities, including Tennis, gym, football, martial arts, swimming and UFC

There has recently been a new skatepark built in Harlow next to Burnt Mill School the project has been funded by investment of over £300,000, largely coming from Harlow Council with £57,500 coming from Sport England. The park also has many security features such as 24/7 CCTV coverage, and is floodlit at night. The 650sq metre park is made entirely from concrete, and has a bowl as well as a street course which contains quarter pipes, flat banks, rails and steps. It is suitable for people of all ages as well as skateboards, inline skates, scooters and BMXs.

Harlow can also lay claim to the 2010 Bowls England Singles Champion when Harlow resident Steve Mitchinson won the final against Scott Edwards from Sussex.

Art and culture

Harlow is the home to a major collection of public sculptures (over 100 in total) by artists ranging from Auguste Rodin to Henry Moore and Barbara Hepworth. Many of these are owned by the Harlow Art Trust, an organisation set up in 1953 by the lead architect of Harlow Frederick Gibberd. Gibberd had idealist notions of the New Town as a place where people who might not normally have access to art could enjoy great sculptures by great artists on every street corner. Consequently almost all of Harlow's sculpture collection is located in the open air, in shopping centres, housing estates and parks around the town.[24]

In 2009 Harlow Council voted to celebrate Harlow's collection of sculpture by branding Harlow as 'Harlow Sculpture Town - The World's First Sculpture Town'. Harlow Sculpture Town began as an initiative from Harlow Art Trust, this will see Harlow present itself to the world as 'Sculpture Town', in a similar way to Hay-on-Wye's presentation of itself as Booktown.[25] [26]

As part of the 'Sculpture Town' branding, Harlow is also home to the Gibberd Garden, the former home of Frederick and Elizabeth Gibberd, which is a managed twentieth-century garden, and home to some of the Gibberd's private sculpture collection.[27]

Harlow is also the location of the Playhouse Theatre,[28] and an art gallery, called the Gibberd Gallery, located in the Civic Centre, containing a collection of twentieth-century watercolours and temporary exhibitions.[29] There are many dance schools in harlow, many of the west end performers trained at the facilities in Harlow.

Environment

A major feature to the new town is its green wedges, with over 1/3 of the town being parkland or open space. Harlow Town Park is one of the largest urban parks in Britain, and occupies a large area of the central town. Each estate is also separated by open space.

The town is in a very dry area of the UK, with nearby Maldon being the driest area in the country. In 2006 the entire South East of England was affected by drought, with Harlow covered by a water restriction order, preventing unnecessary use of water. The area is generally much milder than most other parts of the UK.

The summer of 2006 also saw flash floods hit many parts of the town, causing major roads through the town to become temporarily impassable, and severe damage to many properties around the town. As a result, the council is reviewing its flood defences and drainage systems.

People from Harlow

For a full list, see Category:People from Harlow

Twin towns

- ▙ Havířov, Czech Republic
- ▦ Stavanger, Norway
- ▮▮ Vélizy-Villacoublay, France
- ▩ Tingalpa, Australia

References

[1] Resident Population Estimates (http://www.neighbourhood.statistics.gov.uk/dissemination/LeadTableView.do?a=3&b=276963& c=harlow&d=13&e=13&g=445992&i=1001x1003x1004&m=0&r=1&s=1207489956312&enc=1&dsFamilyId=1812) for Harlow

[2] Harlow District Council (http://www.harlow.gov.uk/about_the_council/council_services/business_services/regeneration_unit/ key_statistics_and_data/harlow_comparison_with_essex.aspx) - Harlow: A Comparison with Essex 2001 Census

[3] Gardens Guide (http://www.gardenvisit.com/b/gibberd.htm) - Frederick Gibberd

[4] New town, a name change and all the jazz (http://news.bbc.co.uk/1/hi/magazine/8603701.stm) BBC News web site

[5] Memorial University - Department of Geography (http://www.mun.ca/geog/interdisiplinary/harlow/harlow1.php) - Harlow's History and Geography

[6] English Heritage - Images of England (http://www.imagesofengland.org.uk/details/default.aspx?pid=1&id=472019) - The Lawn

[7] BBC News (http://news.bbc.co.uk/1/hi/uk/6264579.stm) - *Redeveloping Essex's fallen utopia*

[8] Vision of Britain (http://www.visionofbritain.org.uk/relationships.jsp?u_id=10241649) - Harlow parish

[9] Vision of Britain (http://www.visionofbritain.org.uk/unit_page.jsp?u_id=10241649) - Harlow UD

[10] Ropemaker Properties Limited (http://www.harlownorth.com) - Harlow North

[11] Stop Harlow North Campaign Group (http://www.stopharlownorth.com)

[12] http://umapper.s3.amazonaws.com/maps/kml/45876.kml

[13] "Harlow, First Avenue, Multi-Modal Corridor, Phase 2" (http://www.essexcc.gov.uk/vip8/ecc/ECCWebsite/content/binaries/ documents/Transportation_and_Road_Planning/ Harlow_First_Avenue_-_CIF2_Business_Case_-_Nov_2008_-_Full_Final_Report_-_website_version.pdf?channelOid=null). 2008-11. . Retrieved 2009-10-14.

[14] Mark Hall School (http://www.markhall.essex.sch.uk/)

[15] St Marks School (http://www.st-marks.essex.sch.uk/)

[16] Burnt Mill School (http://www.burntmill.essex.sch.uk)

[17] Stewards School (http://www.stewardsschool.co.uk/)

[18] Passmores Academy (http://www.passmoresacademy.com/)

[19] Harlow College (http://www.harlow-college.ac.uk/)

[20] (http://www.passmoresschool.com/page_viewer.asp?page=New+school&pid=644) - *Information on new school on the Passmores School and Technology College website*. 25 October 2010

[21] Harlow College (http://harlow-college.ac.uk)

[22] http://www.mun.ca/geog/interdisiplinary/harlow/harlow2.php

[23] http://www.harlowpenguins.com/

[24] Gillian Whiteley, *Sculpture in Harlow* (Harlow: Harlow Art Trust, 2005)

[25] Harlow Herald (newspaper), 31 March 2009

[26] http://www.harlowarttrust.org.uk Harlow Art Trust

[27] http://www.thegibberdgarden.co.uk see Gibberd Garden

[28] http://www.playhouseharlow.com The Playhouse

[29] (http://www.harlow.gov.uk/about_the_council/council_services/leisure_and_culture/the_gibberd_gallery.aspx) *harlow.gov.uk*

External links

- Visit Harlow (http://www.visitharlow.com/) - a website from Harlow District Council (http://www.harlow.gov.uk/)

Epping

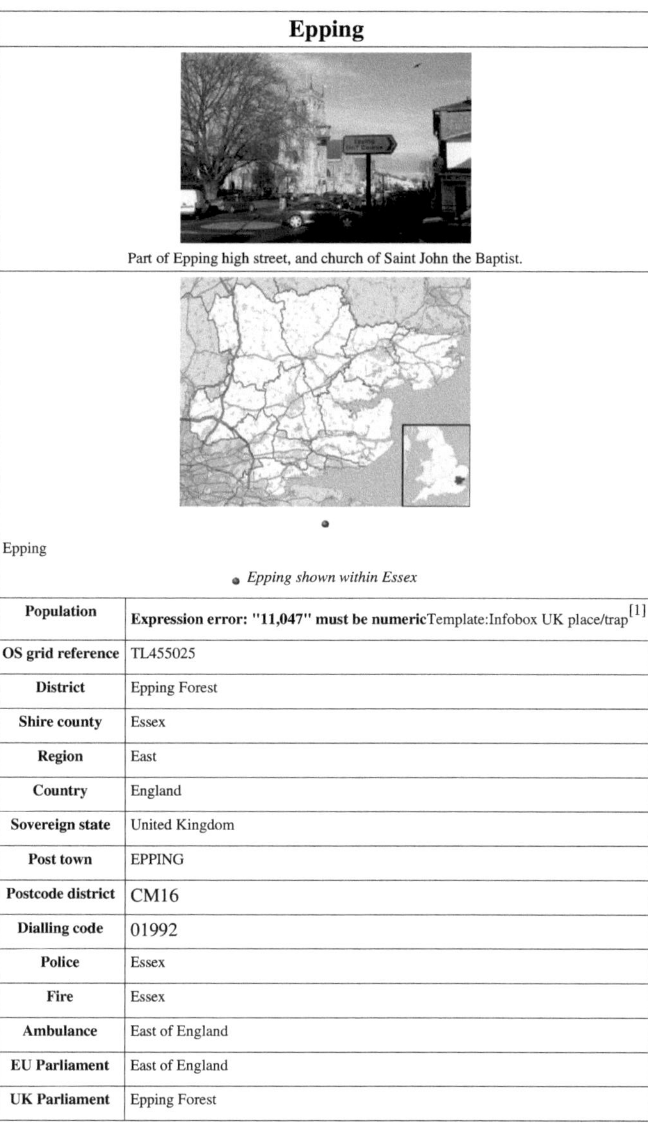

Epping	
Part of Epping high street, and church of Saint John the Baptist.	
Epping *Epping shown within Essex*	
Population	**Expression error: "11,047" must be numeric**Template:Infobox UK place/trap[1]
OS grid reference	TL455025
District	Epping Forest
Shire county	Essex
Region	East
Country	England
Sovereign state	United Kingdom
Post town	EPPING
Postcode district	CM16
Dialling code	01992
Police	Essex
Fire	Essex
Ambulance	East of England
EU Parliament	East of England
UK Parliament	Epping Forest

Website	www.eppingtowncouncil.gov.uk [2]

Epping is a small market town and civil parish in the Epping Forest district of the County of Essex, England. It is located 3.5 miles (**unknown operator: u'strong'** km) north-east of Loughton, 4.6 miles (**unknown operator: u'strong'** km) south of Harlow and 10.9 miles (**unknown operator: u'strong'** km) north-west of Brentwood.

The town retains a rural appearance being surrounded by Epping Forest and working farmland, and has many very old buildings, many of which are Grade I and II listed buildings. The town also retains its weekly market which is held every Monday and dates back to 1253.[3] In 2001 the parish had a population of 11,047[1] although this has increased marginally since then.

Epping has been twinned with the German town of Eppingen in north-west Baden-Württemberg since 1981.[4] Although the once-famous Epping Butter, which was highly sought after in the 18th and 19th centuries, is no longer made, the equally well-known Epping sausages are still manufactured by Church's Butchers who have been trading on the same site since 1888.

History

"Epinga", a small community of a few scattered farms and a chapel on the edge of the forest, is mentioned in the Domesday Book of 1086. However, the settlement referred to is known today as Epping Upland. It is not known for certain when the present day Epping was first settled. By the mid-12th century a settlement known as Epping Heath (later named Epping Street), had developed south of Epping Upland as a result of vigorous clearing of the forest for cultivation. In 1253 King Henry III conveyed the right to hold a weekly market in Epping Street which helped to establish the town as a centre of trade and has continued to the present day (the sale of cattle in the High Street continued until 1961).[5]

The linear village of Epping Heath developed slowly into a small main-road town and by the early 19th century considerable development had taken place along what is now High Street and Hemnall Street. Up to 25 stagecoaches and mailcoaches a day passed through the town from London en route to Norwich, Cambridge and Bury St. Edmunds. In the early 19th century, 26 coaching inns lined the High Street.[6] A couple survive today as public houses, e.g. The George and Dragon and The Black Lion. The advent of the railways put an end to this traffic and the town declined, but it revived after the extension of a branch line from London in 1865 and the coming of the motor car.

A number of listed buildings, most dating from the 18th century, line both sides of the High Street although many were substantially altered internally during the 19th century. Some of the oldest buildings in the town can be found at each end of the Conservation Area, e.g. Beulah Lodge in Lindsey Street (17th century), and the attractive group of 17th and early 18th century cottages numbered 98-110 (even) High Street.[7]

The original parish church, first mentioned in 1177, was All Saints' in Epping Upland, the nave and chancel of which dates from the 13th Century.[8] In 1833, the 14th century chapel of St John the Baptist in the High Road was rebuilt in the gothic style. It became the parish church of Epping in 1888 and was again rebuilt. A large tower was added in 1909.[9]

Today

Epping, as it stands today, has grown as a favoured town of residence for those who work in London. Particularly sought after is the hamlet of Coopersale St where house prices have bucked the national trend and held their values. Its market still brings shoppers in from surrounding villages and towns every Monday. Perhaps the most prominent building in Epping these days is the District Council's office with its clock tower, designed to bring balance to the High Street with the old Gothic water tower at the southern end, built in 1872, and St John's Church tower in the centre. The centre of Epping on and around the High Street is a designated conservation area.[10]

Epping's increasing popularity with young professionals and families, along with the Government's East of England Plan has led to the current situation: Epping is experiencing the biggest threat to its rural status yet and a number of sites (the largest being St. Margaret's Hospital) are being proposed for redevelopment as new housing estates.

The various developments would see Epping's housing stock rise by around 20% and has caused strong opposition from residents who wish to retain Epping's rural 'charm', they state the town does not have the infrastructure to cope with a large influx of new residents and vehicles. Residents point to the regular traffic congestion, lack of parking spaces, low water pressure and total lack of an NHS dentist as examples. This opinion has been echoed by Epping Town Council, who have stated that Epping will not be able to cope with any new housing estates for at least 10 years.[11]

Governance

Epping is part of the Epping Forest parliamentary constituency, represented by Conservative MP Eleanor Laing. From 1924 to 1945, the old Epping division of Essex (which included Woodford, Chingford, Harlow and Loughton as well as Epping) was represented by Winston Churchill. It now sits in the Epping and Theydon Bois division of Essex County Council which is Liberal Democrat held. The town is divided into two district council wards. Epping Hemnall encompasses most of the town south-east of Epping High Street (B1393) including Ivy Chimneys, Fiddlers Hamlet and Coopersale. The rest of Epping lies in Epping Lindsey and Thornwood ward, as does Thornwood in the adjacent parish of North Weald Bassett. Both wards elect three councillors each.

As well as the County and District Councils, Epping has a Town council consisting of 12 councillors, six each elected from Epping Hemnall and Epping Lindsey wards, one of which is elected Mayor of Epping and acts as Chairman of Council, as well as a civic and ceremonial head of the local community.

Epping Forest District Council's headquarters are located in Epping High Street.[12]

Geography

Epping lies 17 miles (**unknown operator: u'strong'** km) north-east of the centre of London towards the northern end of Epping Forest on a ridge of land between the River Roding and River Lea valleys. Epping is north of the small village of Theydon Bois.

Most of the population live in the built up area centred on and around the High Street (B1393) and Station Road. About a thousand people live in the small village of Coopersale which, while physically separated from Epping by forest land, is still part of the civil parish. A few dozen households make up the hamlets of Coopersale Street and Fiddlers Hamlet. Much of the eastern part of the present parish was until 1895 in the parish of Theydon Garnon.

The Town lies north-east of junction 26 (Waltham Abbey, Loughton A121) of the M25 motorway and south-west of junction 7 (Harlow) of the M11 motorway.

Transport

Epping is served by a number of bus routes, serving many surrounding towns and villages including; Thornwood Common, Harlow, Abridge, Waltham Abbey, Ongar and Brentwood. The bus services are either commercial services (routes 7/7A/7B, 19/20/21, 575 and X5), or operated under contract to Essex County Council (routes 213, 381/382, 501 and 541). There are no London Buses services in Epping..

Bus route 541 at Epping Tube Station

Epping is served by Transport for London rail services, and is the eastern terminus of the Central Line of the London Underground. The Central Line now terminates at Epping. However until 30 September 1994, it used to serve stations at North Weald, Blake Hall and Ongar where services terminated. The station has a car park with 508 spaces and is the second largest car park on the London Underground network,[13] a toilet, a ticket machine, a pay phone as well as seats for sitting outside of the station to wait for buses.

Main Line train services are available from a number of neighbouring towns, with the closest stations to Epping being Roydon, Harlow and Chingford, these are served by the West Anglia Main Line and are operated by Greater Anglia. However there is no direct public transport to Roydon and Chingford stations from Epping, making Harlow station the most accessible.

Train

Central Line from Epping Station.

Education

- St John's CE Secondary School, the only secondary school in Epping, is now designated as a specialist Engineering College. The school has an active charity fundraising group led by a Student Executive team. In 2006 two students were awarded the Rotary Prize for 'Service to School' by the local Epping Rotary Club.
- Coopersale Hall School], a private primary school at the end of Flux's Lane, Epping.
- Ivy Chimneys Primary School], a primary school located in Ivy Chimneys, Epping.
- Epping Junior School, a primary school located in the heart of Epping's town centre in St John's Road.
- Coopersale's Primary School (confused with private above not in Coopersale). Founded in 1850 in Coopersale Street, now a house but bell tower still visible, in September 1970 the school was moved into its new premises in the village of Coopersale and has since been called Coopersale and Theydon Garnon C.E. (Vol.Cont.) Primary School.

Sport

Epping Town played in the Isthmian League until folding during the 1984–85 season. Epping FC currently play in the Essex Olympian League. Both have played at Stonards Hill.

Notable residents

- Singer Rod Stewart
- Ex-professional football player and manager Glenn Hoddle
- Actor and television presenter Bradley Walsh
- Actor Nick Berry
- Ex-professional footballer Dennis Rofe
- Former lead singer of rock band Uriah Heep, David Byron
- TV presenter and comedian Griff Rhys Jones
- Television presenter Ben Shephard
- Screenwriter and novelist Julian Mitchell
- Writer and illustrator Jill Barklem
- Politician Philip Hammond
- Supermodel Lisa Snowdon
- Actor James Buckley
- Comedian Alan Davies
- Actress Jessie Wallace
- Band CRASS

Trivia

- Epping's famous weekly market changed from being held every Monday to every Friday from 1575 up until just after the First World War, at which point it returned to being held on Monday.[14]
- Epping is the starting point for the Essex Way, which is a long distance path between Epping and Harwich.[15]
- Epping is home to the first ever Clinton Cards shop which was opened in 1968.[16]

Twin town

Epping is twinned with:

- ▬ Eppingen, Germany[17]

References

- Epping Forest District Council (2005). *Key Facts: 2001 Census* [18].
- *Epping Town Guide*. Plus Publishing Services (on behalf of Epping Town Council. 2002.
- Jenkins (2001). *Churchill*. Macmillan. pp. 391–392. ISBN 0-330-48805-8.
- Parish Profile: Epping [19] - information about Epping from the 2001 census (PDF file)

Sign showing twin towns of Epping

Notes

[1] Parish Profile : Epping (http://www.eppingforestdc.gov.uk/Council_Services/planning/census/Epping.asp)

[2] http://www.eppingtowncouncil.gov.uk

[3] (http://www.eppingtowncouncil.gov.uk/Epping Town Information/Guide 07-08.pdf)

[4] Epping Town Guide (http://www.eppingtowncouncil.gov.uk/Epping Town Information/About.htm)

[5] http://www.british-history.ac.uk/report.aspx?compid=42715

[6] http://www.british-history.ac.uk/report.aspx?compid=42714

[7] EPPING (http://www.eppingforestdc.gov.uk/Council_Services/planning/conservation/EPPING.asp)

[8] http://www.eppinguplandchurch.org.uk/index.php?/histoty-of-the-church.html

[9] http://www.british-history.ac.uk/report.aspx?compid=42716

[10] http://www.eppingforestdc.gov.uk/Library/files/planning/Conservation/Epping.pdf

[11] "EPPING: 'Decade until town is ready for development' (From East London and West Essex Guardian Series)" (http://www.
guardian-series.co.uk/news/3569991.EPPING____Decade_until_town_is_ready_for_development_/). Guardian-series.co.uk.
2008-08-04. . Retrieved 2010-01-05.

[12] information@eppingforestdc.gov.uk. "Epping Forest District Council Home Page" (http://www.eppingforestdc.gov.uk).
Eppingforestdc.gov.uk. . Retrieved 2010-01-05.

[13] Epping station to be refurbished and improved | Transport for London (http://www.tfl.gov.uk/corporate/media/newscentre/3873.aspx)

[14] Epping - Economic history and local government | British History Online (http://www.british-history.ac.uk/report.aspx?compid=42715)

[15] BBC - Essex - Places - Where the town gives way to ancient wood (http://www.bbc.co.uk/essex/content/articles/2006/03/03/
essex_way_day_one_walk.shtml)

[16] "About" (http://www.clintoncards.co.uk/about/). Clinton Cards. . Retrieved 2010-01-05.

[17] "Epping Eppingen Twinning Association - Homepage" (http://www.eppingeppingentwinning.co.uk). Eppingeppingentwinning.co.uk. .
Retrieved 2010-01-05.

[18] http://www.eppingforestdc.gov.uk/Library/files/planning/key%20facts%20-%20final.pdf

[19] http://www.eppingforestdc.gov.uk/Council_Services/planning/census/Epping.asp

External links

- Epping Forest District Council (http://www.eppingforestdc.gov.uk)
- Essex County Council (http://www.essexcc.gov.uk)
- Epping Forest (http://www.cityoflondon.gov.uk/Corporation/living_environment/open_spaces/
epping_forest.htm) - Information about Epping Forest, which is owned and managed by the City of London
Corporation.
- Epping Weather (http://www.eppingweather.co.uk/) - Weather records from the Epping Weather station.
- Epping to Ongar (http://www.urban75.org/london/ongar.html) - Chronicles the history of the now defunct
Ongar branch line.
- Epping Forum (http://www.eppingforum.co.uk/) - Website for Epping residents to discuss local issues.
- Plainly Say No (http://www.psn.weebly.com) - The official website of the campaign against Bellway Homes'
purposed development of 351 homes on part of the St. Margaret's Hospital site.
- Photos of Epping and surrounding area on geograph.org.uk (http://www.geograph.org.uk/search.
php?i=7227190)

Matching_Green

<table>
<tr><th colspan="2" style="text-align:center">Matching Green</th></tr>
<tr><td colspan="2" style="text-align:center">Matching Green

Matching Green shown within Essex</td></tr>
<tr><td>Population</td><td>635 [1]</td></tr>
<tr><td>Civil parish</td><td>Matching</td></tr>
<tr><td>District</td><td>Epping Forest</td></tr>
<tr><td>Shire county</td><td>Essex</td></tr>
<tr><td>Region</td><td>East</td></tr>
<tr><td>Country</td><td>England</td></tr>
<tr><td>Sovereign state</td><td>United Kingdom</td></tr>
<tr><td>Post town</td><td>HARLOW</td></tr>
<tr><td>Postcode district</td><td>CM17</td></tr>
<tr><td>Dialling code</td><td>01279</td></tr>
<tr><td>Police</td><td>Essex</td></tr>
<tr><td>Fire</td><td>Essex</td></tr>
<tr><td>Ambulance</td><td>East of England</td></tr>
<tr><td>EU Parliament</td><td>East of England</td></tr>
<tr><td>UK Parliament</td><td>Brentwood and Ongar</td></tr>
</table>

Matching Green is a village which forms part of the civil parish of Matching, in the County of Essex, England. It is located 3.0 miles (**unknown operator: u'strong'** km) east of Harlow, 4.5 miles (**unknown operator: u'strong'** km) north west of Chipping Ongar and 3.9 miles (**unknown operator: u'strong'** km) south east of Sawbridgeworth.

The village has one of the largest village greens in Essex. Its almost triangular shape extends to 5.6 hectares (13.8 acres) and is lined along each edge by a variety of (mainly) detached cottages and houses ranging in age from the 14th to the 19th century, twenty-eight of which are listed buildings. Since there are relatively few trees, the buildings are important in defining the shape and size of the green.

Matching Parish make-up

- Matching
- **Matching Green**
- Matching Tye

Transport

Bus

Route Number	Route	Operational Details
47 ♿	Harlow to Moreton via Matching Green	Tue-Sat (1 return journey)
147 ♿	Harlow to Ongar via Matching Green	Wed Only (1 return journey)

Also see List of bus routes in Essex

External links

- Matching Parish Council [2]
- The Chequers Pub & Restaurant [3]
- Matching Green Cricket Club [4]
- Matching Green Primary School [5]

References

[1] Parish Profile : Matching (http://www.eppingforestdc.gov.uk/council_services/planning/census/matching.asp). Epping Forest District Council.
[2] http://matchingcouncil.org.uk/index.php
[3] http://www.thechequersmatchinggreen.co.uk/
[4] http://mgcc.hitscricket.com/default.aspx
[5] http://www.matchinggreenprimaryschool.com/

Matching,_Essex

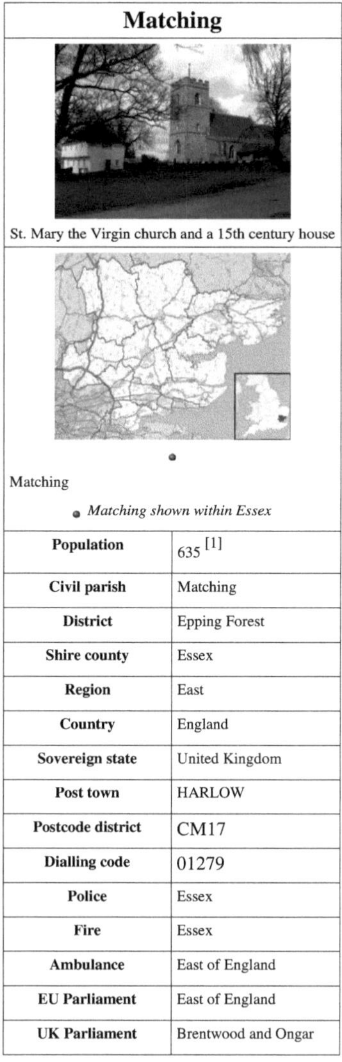

Matching	
St. Mary the Virgin church and a 15th century house	
Matching shown within Essex	
Population	635 [1]
Civil parish	Matching
District	Epping Forest
Shire county	Essex
Region	East
Country	England
Sovereign state	United Kingdom
Post town	HARLOW
Postcode district	CM17
Dialling code	01279
Police	Essex
Fire	Essex
Ambulance	East of England
EU Parliament	East of England
UK Parliament	Brentwood and Ongar

Matching is a village and civil parish in the Epping Forest district of the County of Essex, England.

The village of Matching lying on undulating land is separated from Stort valley by the parishes of Harlow and Sheering. It is a very old settlement and little is known of its early history but the name is of Saxon origin, being derived from the people or tribe of Moecca (Match) who settled in an open area of pasture called an "Ing", hence 'Matching'.

By the time of the Victoria Country History of Essex, Matching had four manor houses standing on or near their medieval sites; namely Matching Hall, Housham Hall, Parvilles and Stock Hall. Watermans Hall is also mentioned but where abouts is obscured, there is a deserted moated site where a manor house could possibly sited at Gunnetts Green but nothing remains there today.

All the owners of Down Hall just north of the church in Hatfield Heath parish played a major part in Matching.

The Church of St Mary the Virgin, there is no mention of this church in the Domesday book but a Norman church was probably built on an old Saxon site. The tower was added in the 15th Century. It is plain, square and embattled and surmounted by a low tiled spire and weather clock. It retains its original 13th Century doorway.

There are 6 bells which were restored in 1990. Originally there were 5 bells in the tower the 6th bell was added in 1887 to celebrate the Jubilee of Queen Victoria. It is inscribed "God Save the Queen". The 2nd and 3rd bells were originally cast about 1500 by William Culverden of Houndsditch and inscribed "Sancte Thoma ora pro nobis" and "Sancta Anna ora pro nobis". Bells number 4 and 6 are inscribed "God Save the King. 1615" and "God Save the King. 1640" They were both made by Robert Oldfield of Hertford.

The clock was removed from the old church at Epping when the church was pulled down and set up in Matching in memory of Henry Selwin-Ibbetson, 1st Baron Rookwood of Down Hall. On the south wall the first window commemorates the restoration of the church by Henry Selwin-Ibbetson, 1st Baron Rookwood and Lady Selwin-Ibbetson later Lord & Lady Rookwood. The other stained window on the south wall is dedicated by parishioners and friends to the memory of Sir Henry Selwin-Ibbetson, 1st Baron Rookwood who died on 15 January 1902. The east window, which commemorates Edan, Lady Rookwood of Down Hall are by Powell of Whitefriars.

The Organ is a rare Bevington with pipe work over the console. A brass plaque commemorates the erection of the orgam by Mrs. Calverley of Down Hall in memory of her brother, Sir Frederick Henniker, of 60th Rifles, who died 19 August 1908.

A memorial plaque remembers the American Airmen who lost their lives in World War II when stationed at Matching. They came from the 391st Bombardment Group of the U.S Ninth Air Force.

A World War II airfield, RAF Matching, was located nearby in Matching Green.

The Welsh poet and clergyman, John Morgan, was curate (1713 −1728) and then vicar (1728−1733 or 1734) here, and as a result gained the nickname "John Morgan Matchin".[2]

Matching Parish make-up

- **Matching**
- Matching Green
- Matching Tye
- Hobbs Cross
- The farm Harlow Tye is sometimes referred to as a hamlet in its own right.

References

[1] Parish Profile : Matching (http://www.eppingforestdc.gov.uk/council_services/planning/census/matching.asp). Epping Forest District Council.

[2] Edwards, Huw M. (2004). "Morgan, John (1688−1733/4)" (http://www.oxforddnb.com/view/article/62912). *Oxford Dictionary of National Biography (online edition, subscription access)*. Oxford University Press. . Retrieved 2008-04-10.

Rik_Mayall

Rik Mayall	
Birth name	Richard Michael Mayall
Born	7 March 1958Harlow, Essex, England
Medium	Television, film, stand up comedy
Nationality	English
Years active	1980–present
Genres	Black comedy, Physical comedy
Influences	Monty Python, Derek and Clive, Rowan Atkinson
Influenced	David Walliams, Matt Lucas, Stewart Lee
Spouse	Barbara Robbin (m. 1985-present, 3 children)
Notable works and roles	*Rick* in The Young Ones *Richie Richard* in Bottom The Comic Strip Presents... *Alan B'stard* in The New Statesman *Richie Twat* in Guest House Paradiso
Emmy Awards	
Outstanding Voice-Over Performance 1997 *The Willows in Winter*	

Richard Michael "Rik" Mayall (born 7 March 1958) is an English comedian, writer and actor. He is known for his comedy partnership with Ade Edmondson, his over-the-top, energetic portrayal of characters, and as a pioneer of alternative comedy in the early 1980s. Such notability has led him to appear in sitcoms such as *The Young Ones*, *Blackadder*, *The New Statesman*, and *Bottom* and even onto the big screen in comedy films such as *Drop Dead Fred* and *Guest House Paradiso*.

Early life

Mayall, the second of four children, was born in Harlow, Essex[1] to John and Gillian Mayall. He has an older brother, Anthony, and two younger sisters, Libby and Kate.[2] When he was three years old, Mayall and his parents — who taught drama — moved to Droitwich Spa, Worcestershire, where he spent the rest of his childhood and performed in his parents' plays. After attending the King's School, Worcester, Mayall went to the Victoria University of Manchester in 1976 to study drama, where he befriended his future comedy partner Ade Edmondson. He also met Ben Elton and Lise Mayer, with whom he later co-wrote *The Young Ones*.

Career

Edmondson and Mayall gained their reputation at the Comedy Store, from 1980. The double act, "20th Century Coyote", became popular. Mayall also developed solo routines using characters such as Kevin Turvey and a pompous anarchist poet named Rick. This led to Edmondson and Mayall, along with Comedy Store compere Alexei Sayle and other upcoming comedians including Nigel Planer, Peter Richardson, French and Saunders, Arnold Brown and Pete Richens, to set up their own comedy club called "The Comic Strip" in the Raymond Revue Bar, a strip club. Mayall's popularity led to a regular slot for Kevin Turvey on *A Kick Up the Eighties*, first broadcast in 1981. He appeared as "Rest Home" Ricky in Richard O'Brien's *Shock Treatment*, sequel to *The Rocky Horror Picture Show*.

He played Dentonvale's resident attendant as the love interest to Nell Campbell's Nurse Ansalong.

Mayall's television appearances as Kevin Turvey in 1977 along with Johnathan PP Seller warranted a mockumentary based on the character entitled *Kevin Turvey — The Man Behind The Green Door*, broadcast in 1982. The previous year, he appeared in a bit role in *An American Werewolf in London*. His stage partnership with Edmondson continued, often appearing together as "The Dangerous Brothers", hapless daredevils whose hyper-violent antics foreshadowed their characters in *Bottom*. Mayall also made a cameo appearance in the 1983 gothic horror film, *The Keep* directed by Michael Mann. Channel 4 offered the Comic Strip group six short films, which became *the Comic Strip Presents...*, debuting on 2 November 1982. The series, which continued sporadically for many years, saw Mayall play a wide variety of roles. It was known for anti-establishment humour and for parodies such as *Bad News On Tour*, a spoof "rockumentary" starring Mayall, Richardson, Edmondson and Planer as a heavy metal band.

At the time *The Comic Strip Presents...* was negotiated, the BBC took an interest in *The Young Ones*, a sitcom written by Mayall and then-girlfriend Lise Mayer, in the same anarchic vein as *Comic Strip*. Ben Elton joined the writers. The series was commissioned and first broadcast in 1982, shortly before *Comic Strip*. Mayall played Rik, a pompous sociology student and Cliff Richard devotee. Despite the sitcom format, Mayall maintained his double-act with Edmondson, who starred as violent punk Vyvyan. Nigel Planer (as hippie Neil) and Christopher Ryan (as "Mike the cool person") also starred, with additional material written and performed by Alexei Sayle. The first series was successful and a second was commissioned in 1984. The show owed a comic debt to Spike Milligan, but Milligan was disapproving of Mayall, and once wrote: "Rik Mayall is putrid - absolutely vile. He thinks nose-picking is funny and farting and all that. He is the arsehole of British comedy." [3]

Becoming a household name

Mayall continued to work on *The Comic Strip* films. He returned to standup, starring on *Saturday Live* — a British version of the American *Saturday Night Live* — first broadcast in 1985. He and Edmondson had a regular section as "The Dangerous Brothers", their earlier stage act. In 1985, Mayall debuted another comic creation. He had starred in the final episode of *The Black Adder* in 1983 as "Mad Gerald". He returned to play Lord Flashheart in the *Blackadder II*-episode entitled "Bells". A descendant of this character, Squadron Commander Flashheart, was in the *Blackadder Goes Forth* episode "Private Plane". In the same episode, he was reunited with Edmonson, who played German flying ace Baron von Richtofen the "Red Baron", in a scene where he comes to rescue Captain Blackadder from the Germans. Nearly a decade later, Mayall also appeared in *Blackadder: Back & Forth* as Robin Hood.

In 1986, Mayall joined with Planer, Edmondson and Elton to star in *Filthy Rich & Catflap* as Richie Rich in what was billed as a follow-up to *The Young Ones*. The idea of "Filthy Rich and Catflap" was in reaction to comments Jimmy Tarbuck made about the "Young Ones". The series primary focus was to highlight the "has been" status of light entertainment. While Mayall received positive critical reviews, viewing figures were poor and the series was never repeated on the BBC. In later years, release on video, DVD and repeats on UK TV found a following. Mayall suggested the series did not last because he was uncomfortable acting in an Elton project, when they had been co-writers on *The Young Ones*.[4] 1987 saw Mayall co-star with Edmondson in the ITV sit-com *Hardwicke House*. Due to adverse reaction of press and viewers, ITV withdrew the series after two episodes.[5] The same year, Mayall had a number one hit in the UK Singles charts when he and his co-stars from *The Young Ones* teamed with Cliff Richard to record "Living Doll" for the inaugural *Comic Relief* campaign. Mayall played Rick one last time in the stage show and has supported the *Comic Relief* cause ever since. He appeared on the children's television series *Jackanory*. His crazed portrayal of Roald Dahl's *George's Marvellous Medicine* proved memorable.[6] However, the BBC received complaints "with viewers claiming both story and presentation to be both dangerous and offensive."[7]

In 1987, Mayall played fictional Conservative MP Alan Beresford B'Stard in the sitcom *The New Statesman* for Yorkshire Television, written by Laurence Marks and Maurice Gran. The character was a satire of Tory MPs in the United Kingdom in the 1980s and early 1990s. The programme ran for four series — incorporating two BBC specials — between 1987–1994 and was a success critically and in ratings. In a similar vein to his appearance on

Jackanory, in 1989, Mayall starred in a series of bit shows for ITV called *Grim Tales*, in which he narrated Grimm Brothers fairy tales while puppets acted the stories. In the early 1990s Mayall starred in humorous adverts for Nintendo games and consoles. With money from the ads, he bought his house in London which he calls "Nintendo Towers".

1990s

Adrian Edmondson (left) and Rik Mayall (right) as Eddie and Richie in *Bottom*

In 1991, Edmondson and Mayall co-starred in the West End production of Beckett's *Waiting for Godot* at the Queen's Theatre. Here they came up with the idea for *Bottom*, which they said was a cruder cousin to *Waiting for Godot*. *Bottom* was commissioned by the BBC and three series were shown between 1991–1995. Mayall starred as "Richard 'Richie' Richard" alongside Edmondson's "Eddie Elizabeth Hitler". The series featured slapstick violence taken to new extremes. The series gained a strong cult following. In 1993, following the second series, Mayall and Edmondson decided to take a stage show version of the series on a national tour. *Bottom: Live* was a commercial success, filling large venues. Four additional stage shows were embarked upon in 1995, 1997, 2001 and 2003, each to great success. The violent natures of these shows saw both Edmondson and Mayall ending up in hospital at various points. A film version, *Guest House Paradiso*, was released in 1999. A fourth TV series was also written, but not commissioned by the BBC.

Mayall starred alongside Phoebe Cates in 1991's *Drop Dead Fred* as the eponymous character, a troublesome imaginary friend reappearing from a woman's childhood. He also appeared in *Carry On Columbus* (1992) with other alternative comedians. In 1991 he played Vladimir in Samuel Beckett's *Waiting for Godot* at the Queen's Theatre in the West End, alongside Edmondson (Estragon) and Christopher Ryan (Lucky). Mayall also provided the voice of the character Froglip, the leader of the goblins, in the 1992 animated film adaption of the 1872 children's tale, *The Princess and the Goblin* by George MacDonald. In 1993, he appeared in *Rik Mayall Presents*, three individual comedy dramas. Mayall's performances won a Best Comedy Performer award at that year's British Comedy Awards, and

Phoebe Cates (left) and Mayall as Elizabeth and Fred in *Drop Dead Fred*.

a second series of three was broadcast in early 1995. He provided the voice for Little Sod in Simon Brett's *How to Be a Little Sod*, written in 1991 and adapted as 10 consecutive episodes broadcast on the BBC in 1995. In the early 1990s, he auditioned for the roles of Banzai, Zazu and Timon in *The Lion King*. He was asked to audition by lyricist Tim Rice. The role of Zazu finally went to his *Blackadder* co-star Rowan Atkinson.

In 1995 Mayall co-starred in a production of the play *Cell Mates*, alongside Stephen Fry. Not long into the run, Fry had a nervous breakdown and fled to Belgium, where he remained for several days, and the play closed. In 2007, Mayall said of the incident: "You don't leave the trenches ... selfishness is one thing, being a cunt is another. I mustn't start that war again."[8] Edmondson poked fun at the event during their stage tours. In *Bottom Live: The Big Number Two Tour*, after Mayall gave mocking gestures to the audience and insulted their town in a silly voice, Edmondson said "Have you finished yet? It's just I'm beginning to understand why Stephen Fry fucked off." In *Bottom Live 2003: Weapons Grade Y-Fronts Tour*, after Richie accidentally fondles Eddie, he replies "I see why Stephen Fry left that play." Towards the end of *Cell Mates* Mayall revealed a replica gun — a prop from the play —

to a passer-by in the street. He was cautioned over the incident. Mayall later conceded that this was "incredibly stupid, even by my standards".[9] Since 1999, Mayall was the voice of the black-headed seagull Kehaar in the first and the second season of the animated television series *Watership Down*.

2000s–present

In 2000, Mayall lent his voice to the PlayStation and Windows PC video game *Hogs of War*. Also that year, Mayall appeared in the video production of *Jesus Christ Superstar* as King Herod. He joked in the "making of" documentary, which was included on the DVD release, that "the real reason why millions of people want to come and see this is because I'm in it! Me and Jesus!" In 2002, Mayall teamed up with Marks and Gran once more when he starred as Professor Adonis Cnut in the ITV sitcom, *Believe Nothing*. However, the sitcom failed to repeat the success of *The New Statesman* and lasted only one series. Following 2003's *Bottom: Live* tour, *Bottom 5: Weapons Grade Y-Fronts*, Mayall stated that he and Edmondson would return with another tour.[10] Shortly thereafter, however, Edmondson told *The Daily Mail* that he no longer wished to work on *Bottom*. This effectively dissolved their nearly 30-year partnership. Edmondson claimed they were "too old" to continue portraying the characters. Edmondson added that, since Mayall had recovered from his coma, he was slower on the uptake and it had become more difficult to work with him, as well as citing that due to taking medication, Mayall had been advised to stop drinking alcohol. However, Edmondson said that the pair remained very close friends.[11]

Mayall voiced Edwin in the BBC show *Shoebox Zoo*. He released an 'in-character' semi-fictionalised autobiography in September 2005 entitled *Bigger than Hitler, Better than Christ* (ISBN 0-00-720727-1). At the same time, he starred in a new series for ITV, *All About George*. Mayall reprised the role of Alan B'Stard in 2006 in the play *The New Statesman 2006: Blair B'stard Project*, written by Marks and Gran. By this time B'Stard had left the floundering Conservatives and become a Labour MP. Following a successful two-month run in London's West End at The Trafalgar Studios in 2007, a heavily re-written version toured theatres nationwide, with Marks and Gran constantly updating the script to keep it topical. However, Mayall succumbed to chronic fatigue and flu in May 2007, and withdrew from the show. Alan B'Stard was played by his understudy, Mike Sherman during his hiatus.

Mayall was cast as the poltergeist Peeves in *Harry Potter and the Philosopher's Stone*, the first of the Harry Potter films in 2001.[12] He claimed in his semi-autobiographical book "Bigger than Hitler, Better than Christ" that he had not been made aware that his scenes had been cut until the full film was officially unveiled at the premiere. He tells the story of this hiring/firing on his second website blog for his 2008 film, *Evil Calls: The Raven*. For *Evil Calls*, he shot his role as Winston the Butler in 2002, when the film was titled *Alone in the Dark*. The film was not completed until 2008 and was released under its new 'Evil Calls' title to distance itself from the *Alone in the Dark* computer game movie. He may appear in a possible sequel. Mayall provides the voice of the Andrex puppy in the UK TV commercials for Andrex toilet paper, and also has a voice part in the UK Domestos cleaning product adverts.[13] He performs the voice of King Arthur in the children's television cartoon series *King Arthur's Disasters*, alongside Matt Lucas (Little Britain) who plays Merlin. Mayall also had a recurring role in the Channel Five remake of the lighthearted drama series, *Minder*.

In September 2009, Mayall played a supporting role in the British television program *Midsomer Murders*, shown on ITV1 (and made by Meridian Broadcasting) as 'David Roper', a recovering party animal and tenuous friend of the families in and around Chettham Park House.

In April 2010, Motivation Records released[14] Mayall's England Football anthem 'Noble England' for the 2010 FIFA World Cup which he recorded with Coventry producer Dave Loughran [15],[16] On the track Mayall performs an adapted speech from Shakespeare's Henry V [17] In June 2010 the official BBC Match Of The Day compilation CD (2010 Edition) was released by Sony/Universal featuring Noble England - Track 18, CD2.

In September 2010 *Cutey And The Sofaguard* was released as an audio book, narrated by Mayall. The book was written by Chris Wade and released by Wisdom Twins Books as a digital download. Mayall told Hound Dawg Magazine that he had the task of voicing 25 different characters. In this same month Mayall played the voice of

Roy's Dad in cult animation *Dog Judo*. He recorded five episodes all of which exclusively streamed from http://www.dogjudo.com.

In November 2010, Mayall provided narrative for five characters for CDs accompanying books from the publishing house Clickety Books [18]. The books are for children and aid speech and language development by bombarding the child with troublesome sound targets.

Rik has so far recorded introductions and narratives for the titles:Jake the Achy Snake [19], Corky the Squawky Hawk [20], Erica the Picky Chicken [21], Lucky the Plucky Duck [22] and Jacqueline the Black Alpaca [23]. The books are written by Craig Green and illustrated by Sarah-Leigh Wills. The recording and prouction was undertaken by Matt Bernard and Jay Auborn at the Deep Blue Sound [24] studio in Plymouth, UK.

On 5 March 2011, Mayall appeared on *Let's Dance For Comic Relief* in which he came on stage and attacked Ade Edmondson with a frying pan during his performance of *The Dying Swan* ballet. Edmondson mentioned backstage that it was the first time in eight years they've done something like that together and claimed Mayall had left his head with a small bump.

In April 2011, Mayall again revived the character of Alan B'Stard to make an appearance in a satirical television advertisement for the No2AV campaign prior to the 2011 voting reform referendum in the UK. The character is shown being elected under the alternative vote system, then using his newly gained position of power to renege on his campaign promises. In his personal life, Rik Mayall does not support the alternative vote. In May of the same year Mayall became the eponymous 'Bombardier' in a TV advertising campaign for Bombardier Bitter in the UK, arguably reviving his Lord Flashheart persona from the UK TV series *Blackadder*.[25]

Personal life

Family

Since 1985 Mayall has been married to Scottish make-up artist Barbara Robbin, with whom he has three children: Rosie (born 1986), Sidney (born 1988) and Bonnie (born 18 September 1995). The couple met in 1981 while filming *A Kick Up The Eighties*. At the time, Mayall was in a long-term relationship with Lise Mayer. Mayall and Robbin embarked on a secret affair which lasted until 1985. Mayall and Robbin immediately eloped to Barbados. Mayer would later suffer a miscarriage.[26] Mayall maintains that, despite a longstanding feud, he and Mayer are now friends.[26]

Quad bike accident

On 9 April 1998, Mayall was injured after crashing a quad bike near his home in Devon.[27] Mayall's daughter Bonnie and her cousin had asked him to take them for a ride on the bike — a Christmas gift from his wife — but he refused due to rain, and went alone. Mayall's wife Barbara looked out the window and saw him lying on the ground with the bike, believing he was joking. He was in a coma for several days. Mayall was airlifted to Plymouth's Derriford Hospital, with two hematomas and a fractured skull. During the following 96 hours, Mayall was kept sedated to prevent movement which could cause pressure on his brain. His family was warned he could die or have brain damage.[28]

After five days doctors felt it safe to bring Mayall back to consciousness. In his 2005 spoof biography, Mayall claims he "rose from the dead". During Mayall's hospitalisation, the *Comic Strip* special *Four Men in a Car* was broadcast for the first time. The film involves Mayall's character being hit by a car. Mayall believed he was held hostage at the hospital. After transfer to hospital in London, he took a taxi home but was taken back that day after being sedated. He was to take medication for a year to prevent epileptic seizures, but stopped taking it, which resulted in him having several epileptic seizures, during one of which he bit through his tongue. He is now on medication for life. Mayall returned to work with voice-over work. His first post-accident job was in the 1998 *Jonathan Creek* Christmas special, as DI Gideon Pryke. Mayall and Edmondson have joked about this event in stage versions of *Bottom*,

Edmondson quipping: 'If only I'd fixed those brakes properly'. The pair wrote the first draft of their feature film *Guest House Paradiso* while Mayall was hospitalised. They planned to co-direct but Edmondson took on the duties himself.

Recognition

In the 2005 poll *The Comedians' Comedian*, Mayall was voted among the top 50 comedy performers of all time.

In 2008, Mayall was awarded an honorary Doctor of Letters (DLitt) from the University of Exeter.[29]

In the 2010 poll "Top 100 Stand-Up Comedians", Mayall was placed 91.

Filmography

Television

Year	Title	Role	Notes
1982	*Whoops Apocalypse*	Biff	Episode: "Autumn Cannibalism"
1982–1984	*The Young Ones*	Rick	2 series
1983	*The Black Adder*	Mad Gerald	Episode: "The Black Seal"
1983–2011	*The Comic Strip Presents...*	Various roles	Several episodes and specials (appeares in 18 of the 40 episodes)
1985	*Happy Families*	Priest	Episode: "Madeleine"
1986	*Saturday Live*	Richard Dangerous	Sketches featuring The Dangerous Brothers
1986	*Blackadder II*	Lord Flashheart	Episode: "Bells"
1987	*Filthy Rich & Catflap*	Gertrude "Richie" Rich	1 series
1987–1994	*The New Statesman*	Alan Beresford B'Stard	4 series
1989	*Blackadder Goes Forth*	Lord Flashheart	Episode: "Private Plane"
1991–1995	*Bottom*	Richard "Richie" Richard	3 series
1995	*The World of Peter Rabbit and Friends*	Tom Thumb (voice)	Episode: "The Tale of Two Bad Mice and Johnny Town-Mouse"
1997	*The Canterville Ghost*	Reverend Dampier	TV film
1998	*Jonathan Creek*	Pryke	Episode: "Black Canary" (Christmas Special)
1999	*Blackadder: Back & Forth*	Robin Hood	TV special (one episode)
1999	*Watership Down*	Keeyar (voice)	Series 1 and 2 (of 3) only
2002	*Believe Nothing*	Quadruple Professor Adonis Cnut	1 series
2009	*Agatha Christie's Marple*	Alec Nicholson	Episode: "Why Didn't They Ask Evans?"
2009	*Midsomer Murders*	David Roper	Episode: "The Creeper"

Film

Year	Title	Role	Notes
1981	*Eye of the Needle*	Sailor	
1981	*An American Werewolf in London*	Man in Pub	
1981	*Shock Treatment*	"Rest Home" Ricky	
1986	*Whoops Apocalypse*	Specialist Catering Commander	
1987	*Eat the Rich*	Micky	Feature film from The Comic Strip Presents...
1991	*Drop Dead Fred*	Drop Dead Fred	
1991	*Little Noises*	Mathias	
1991	*The Princess and the Goblin*	Prince Froglip (voice)	Dubbed voice for the 1992 English-language version
1992	*Carry On Columbus*	The Sultan	
1995	*The Snow Queen*	The Robber King (voice)	
1997	*Bring Me the Head of Mavis Davis*	Marty Starr	
1999	*Guest House Paradiso*	Richard Twat	
1999	*A Monkey's Tale*	Gerard the Gormless (voice)	Dubbed voice for the 2000 English-language version
2001	*Kevin of the North* (aka *Chilly Dogs*)	Carter	
2004	*Churchill: The Hollywood Years*	Baxter	
2005	*Valiant*	Cufflingk (voice)	
2012	*Eldorado*	Chef Mario	

Stage

Year	Title	Role	Notes
1993	*Bottom Live*	Richard "Richie" Richard	Recorded at the Mayflower Theatre in Southampton
1995	*Bottom Live: The Big Number Two Tour*	Richard "Richie" Richard	Recorded at the New Theatre in Oxford
1997	*Bottom Live 3: Hooligan's Island*	Richard "Richie" Richard	Recorded at the Hippodrome in Bristol
2001	*Bottom Live 2001: An Arse Oddity*	Richard "Richie" Richard	Recorded at the Royal Concert Hall in Nottingham
2003	*Bottom Live 2003: Weapons Grade Y-Fronts Tour*	Richard "Richie" Richard	Recorded at the Cliffs Pavilion in Southend-on-Sea

References

[1] NNDB biodata (http://www.nndb.com/people/809/000060629/)

[2] Rik Mayall Biography in *Film Reference* (http://www.filmreference.com/film/96/Rik-Mayall.html)

[3] From a 2002 BBC Obituary of Spike Milligan, for which the author has been unable to find the original, cited in No Such Thing as Society by Andy McSmith, Constable 2011, page 15 ISBN 978-1-84901-979-8

[4] Mayall, interviewed for a *Comedy Connections* profile of *The Young Ones*

[5] UK Online website (http://web.ukonline.co.uk/sotcaa/sotcaa.html?/sotcaa/archive/hardwicke.html)

[6] Acknowledged here (http://news.bbc.co.uk/1/hi/entertainment/tv_and_radio/4648357.stm) on the BBC News website

[7] Reported at Television Heaven website (http://www.televisionheaven.co.uk/jackanory.htm)

[8] Interview with Theatre.com 11 January 2007 (http://www.orangeneko.com/Rik/library/theatre111.htm)

[9] Police Rebuke Rik Mayall for 'Stupid' Gun Prank (http://www.orangeneko.com/Rik/library/telegra8.htm)

[10] The Rik Mayall FAQ (http://www.orangeneko.com/Rik/faq/tourbottom.htm)

[11] Article in *The Daily Mail*'s "Weekend" supplement (2003)

[12] "Rik Mayall Acts Up" (http://www.orangenko.com/Rik/library/iccroydon.htm)

[13] House of Fear.co.uk (http://www.houseoffear.co.uk/films/raven1/raven1.html)

[14] Noble Englands Record Label (http://www.motivationrecords.co.uk).

[15] http://www.daveloughran.com

[16] The Producer of Riks Football Anthem (http://www.daveloughran.com)

[17] Noble England — Riks Football Anthem (http://www.nobleengland.com)

[18] http://www.clicketybooks.co.uk

[19] http://www.clicketybooks.co.uk/shop/jake-the-achy-snake/c-005-jas/

[20] http://www.clicketybooks.co.uk/shop/corky-the-squawky-hawk/c-002-csh/

[21] http://www.clicketybooks.co.uk/shop/erica-the-picky-chicken/c-003-epc/

[22] http://www.clicketybooks.co.uk/shop/lucky-the-plucky-duck/c-004-lpd/

[23] http://www.clicketybooks.co.uk/shop/jacqueline-the-black-alpaca/c-001-jpa/

[24] http://www.deepbluesound.co.uk/studio/

[25] (http://www.thedrum.co.uk/news/2011/05/13/21475-rik-mayall-stars-in-4m-campaign-for-bombardier-beer/)

[26] Interview with Roz Laws, IC Birmingham, 29 December 2002

[27] http://news.bbc.co.uk/1/hi/uk/77157.stm

[28] Interviewed by Michael Owen, *You*, 21 November 1999

[29] http://www.exeter.ac.uk/honorarygraduates/2009/prev_hongrads.shtml

External links

- Rik Mayall (http://www.imdb.com/name/nm562201/) at the Internet Movie Database
- Rik Mayall (http://www.screenonline.org.uk/people/id/476762) at the British Film Institute's Screenonline
- Interview with Rik Mayall on Theatre.com (http://www.theatre.com/story/id/3005429/)
- (http://www.scribd.com/doc/37736644/Hound-Dawg-Issue-10-RIK-MAYALL-INTERVIEW)

Article Sources and Contributors

Matching_Tye *Source*: http://en.wikipedia.org/w/index.php?title=Matching_Tye *Contributors*: Felix Folio Secundus, Maurice45, Mjroots, Redrose64, Spymo, Stephenb, Stepheng3, 11 anonymous edits

Sawbridgeworth *Source*: http://en.wikipedia.org/w/index.php?title=Sawbridgeworth *Contributors*: A.Syed92, Ab65, Acalamari, After Midnight, Alabama Moon, Alex12341000, Andybruce69, Babylon77, Bazj, C2r, CALR, Coolhawks88, Crusoe8181, Dallan72, Danno uk, David Edgar, Debresser, Dogsnuts, DominicConnor, Drpickem, Dubmill, Dylanmills, Ettrig, FJPB, Friend of the Facts, Griffinofwales, Huw4beynon, Iainh, Jeremy Bolwell, JohnAlbertRigali, Josemanimala, Jpbowen, Kev Price, Lifebaka, LilHelpa, Lozleader, Lupin, MRSC, Marek69, Markdmason, Marnanel, Mauls, Mhvn, Myopic Bookworm, Nedrutland, Neutrality, Orioane, Pdean52, Pit-yacker, Regan123, Rojomoke, Sirex98, SlamDiego, Smb1001, Stephenb, Terracescot, TheParanoidOne, Tmol42, Trident13, Velella, Warofdreams, Wayward, Wereon, 120 anonymous edits

Essex *Source*: http://en.wikipedia.org/w/index.php?title=Essex *Contributors*: 2spk, 96.149, A bit iffy, Acalamari, Accurizer, Acknowledged74, Adlea111, Agnetha, Alansohn, Alexneal94, Alexporter vdc, Alfie66, Alxeedo, Amazonien, Amovrvs, Andre Engels, Andre3004, Andrew Dalby, Andrew Norman, Andylee, Angela, Anna Lincoln, Anwar saadat, Askbuster, Aslam nathoo, AssegaiAli, AuburnPilot, Avoided, B21b16, BD2412, BRG, Balloonguy, Barnabypage, BazookaJoe, Bcp67, Beachy, Ben pryer, Betacommand, Bfinn, Biglovinb, Biker Biker, BillTenner, Blanchardb, Bluemoose, BokicaK, Bongwarrior, BrownHairedGirl, Bwithh, CambridgeBayWeather, Can't sleep, clown will eat me, Cancun771, Caomhin, CapitalR, Caponer, Chanheigeorge, Charles01, Cheeesemonger, Chichodidu, Chris j wood, Chrisieboy, Cmbedrock, Cnyborg, Conversion script, Costesseyboy, Cpl Syx, DFS454, Darianthomson, DarkFalls, DaveK@BTC, Dbachmann, Deb, Deflective, Derek Ross, Digitalblues, DinosaursLoveExistence, DominicConnor, Donmike10, DougsTech, Dpaajones, Dr Greg, DragonflySixtyseven, Dukesuns, Dumelow, Dunc1971, Dusimpson, Dynamite Riley, EdBever, Edgar181, EeepEeep, Eichendorff51, El C, Enjoyuk, Enteryourmum, Eopsid, Epbr123, Erianna, EricITOworld, Essex boy, Essexman, Ewx, Excirial, Exok, Fallschirmjäger, Fenners, Filanca, Flatterworld, Floridasand, Footings, Francium12, Francs2000, Fratrep, Fyyer, Fæ, Gaius Cornelius, Gdlimelight, Gdm, GeoManikz, Glen D Dryhurst, Gogo Dodo, Gr8opinionater, Graham87, Gregoryj77, Grimlund, Guinness, Gwinglegog, HOBROUGH, Hadal, Hagedis, Hayden120, Heenan73, Heimstern, HelioSmith, Henrysaunders2, Heron, IJMacD, IW.HG, Iamthehorse, Ian Dalziel, IkbenFrank, Ilikeeatingwaffles, Ipsingh, Iridescent, Iulianu, JaJaWa, Jammydodger, Javert, JayKeaton, Jayunderscorezero, Jeff G., Jeni, JeremyA, JiFish, Jim1138, Jj137, Jncraton, Joevsimp, John C Willis, John F Ryall, John of Reading, Johnm young, Jvhertum, Jx-10, Jza84, Katieh5584, Kaygrub, Kbthompson, Kchishol1970, Keith D, Keith Edkins, Kenneth Shabby, Kf4bdy, Kingboyk, KirkEN, Kkgg101, Koavf, Ksbrown, Kummi, Kurtle, Kwamikagami, LadyofShalott, Le Creillois, Leepinglemur, Lezek, Libby norman, Lightmouse, LinuxDude, Loren.wilton, Lost tourist, Lubbutton, Lukeegan07, Luna Santin, Lundgren8, MER-C, MJCdetroit, MRSC, Mackan, Mahahahaneapneap, Mais oui!, Malcolm Farmer, Malxirocon, Marek69, Mark J, MarkGallagher, Marknew, Marnanel, Matt Brennen, Matticus78, Mauls, Maurice45, Max Naylor, Mcarmichael, Mel Etitis, Mjrogers50, Montgomery '39, Monty845, Morwen, Motmit, MrWeeble, Nakon, Nashikawa, Naturenet, Nedrutland, Nevilley, Newchauncey, Newsquestpaul, Nick C, Nikai, Nilfanion, Nk, Nockely, Nono64, Northmetpit, Nottmlad, Oda Mari, Ohconfucius, Old Moonraker, OldSpot61, Oliver Chettle, Oneblackline, Our Phellap, Owain, Owisped, Palica, Paronomasia, Paul W, Paul-L, Pchurch, Pearle, Peter Shearan, PeterEastern, Philip Stevens, Pinethicket, Plucas58, Pmanderson, Pocketsizepete, Possum, Pterre, RMHED, RachelBrown, Ramdomer, RedWolf, Regan123, Reinthal, Retew Boy, RexNL, Rich Farmbrough, Ripshack, Rjwilmsi, RobertG, Robertpeel1, RocknRollFan21, Romit3, RoyBoy, Rrius, Rror, S19991002, SE7, SFC9394, SP-KP, Saga City, Sam Korn, Sansonic, Serein (renamed because of SUL), Sfan00 IMG, Shawisland, Sheogorath, Simon 0690, Simondaw, Sjc, Skinnr, Slatersteven, Slightsmile, Smalljim, Snigbrook, Snowmanradio, Spellcast, SpuriousQ, Stan Shebs, Stephenb, Student7, Syrthiss, Tarquin, Taxico, Template namespace initialisation script, Terryjoyce, Thatguy820, TheMadBaron, Themfromspace, Thomas888b, Thumperward, TicketMan, Tillypatterson, Timeout123, Timrollpickering, Tiyoringo, TomPhil, TonyW, Tri6ky, Troglopedetes, Tyoungd, Tzartzam, Uakari, Ugal1996, UnQuébécois, Uncle Dick, Useight, Vegaswikian, Velella, Vervin, WOSlinker, Waggers, Warofdreams, Wavelength, Whohe!, William Avery, Woohookitty, Wotnow, Xezbeth, Xkaylesx, 625 anonymous edits

Harlow *Source*: http://en.wikipedia.org/w/index.php?title=Harlow *Contributors*: 4twenty42o, Acalamari, After Midnight, Ahpook, Al Silonov, Alansohn, Aldux, Alex12341000, AndrewN, Andy, Andy Smith, Andybruce69, Artcyprus, AshleyT211, Balbs, BannyFT, Beerandbiscuits, Benjamminsanders, Bhadani, Blue.Scribe, Bobo192, Brett7boi, Brianyoumans, BritishBC, BrownHairedGirl, CWY2190, Cactus.man, Callum, CambridgeBayWeather, Can't sleep, clown will eat me, Caomhin, Carmelite, CarolGray, Charlesdrakew, Chilkoot, Chris4uk, Chrisclem, Cnyborg, Cometstyles, Courcelles, Crumpetman, Cst17, Custardninja, DJH, Darwinek, Davewild, David Edgar, DavidLevinson, Dazhaz, Del5300, Deor, Discospinster, Djuneyt tr, Dkostic, Domjohn, Dr. A. McBrandy, Earl2, Edward, Edward321, Electrictrousers, Elemesh, EricITOworld, EricSerge, Erik9, EvocativeIntrigue, F1Krazy, Famousbec, FrFintonStack, Francs2000, Fratrep, G-Man, G51, Gatorzzz, Geedubber, George2001hi, Giler, Giraffedata, Gr8opinionater, Greebowarrior, Grosenvinge97, Ground Zero, Grstain, Hailey C. Shannon, Happysailor, Hinesh, Huge golf fan, I'MSKYHiGH, Iohannes Animosus, Iridescent, Irons1990, IsarSteve, J.delanoy, JForget, JHoltzman, JLaTondre, James Brown, Jamestaylor, JamieS93, Jamsta, Jdforrester, Jeremy Bolwell, JeremyA, Jhamez84, JiFish, Jimbledon, JoeBamber16, Jordan73933, Jporter1983, JustAGal, JusticeMondeo, KathrynLybarger, Keith Edkins, Kentem, Keresaspa, Kevinhuse, Kierant, King Pickle, Kingkong77, Klemen Kocjancic, Klomphy, Ksbrown, LGreen91, Lakester10, Laurel Bush, Lawrence Cohen, Lee Gregz, Lilm2k9, Little Miss Teacher, LittleHM, Lupin, Luwilt, MRSC, Mais oui!, Mandarax, Manyenemy, Manzanog, MapsMan, Marek69, Markking, Marnanel, Martarius, MartinUK, Mavic Chen, Meklin, Merotoker1, Michael Bednarek, Mindmatrix, Morsus, MortimerCat, Morwen, Mr Larrington, Nilfanion, Nonky, Normanstrike, Oliver Chettle, Oneblackline, Orioane, OwenBlacker, Pegsi, Philippe, Pjburland, Plingsby, Poppunkluva, Project FMF, QuickClown, RFBailey, Readallaboutit, Regan123, RexNL, RobertG, Robertvan1, Robina Fox, Rock Soldier, Rojomoke, Ronhjones, Rooney the waster, Rrose Selavy, Ruy Pugliesi, Saga City, Scissorkicks, Secretlondon, Shinmawa, Signalhead, Simple Bob, Simple Pieman, Simply south, Snowmanradio, Spymo, Srikeit, Stephenb, Strathdon, TBall84, Tassedethe, Template namespace initialisation script, Teskey@btinternet.com, The Anome, The Thing That Should Not Be, Thumperward, Thunderlightning, Tiddly Tom, Tide rolls, Tim1357, Timeineurope, Tiptoety, TobascoSauce, Tower1965, Tresiden, Trident13, Tubecopy2, Tungol, Ukbulla, UrgentFusionguy, Violetbeau, Volatile, Voyageresu, Vrenator, Vulcan's Forge, Waggers, Warofdreams, Was Once, Wereon, White Shadows, White1928, Will-h, William Avery, Winston365, Wipsenade, Zalgo, Zink Dawg, Zzzzzzzzzz1234, 511 anonymous edits

Epping *Source*: http://en.wikipedia.org/w/index.php?title=Epping *Contributors*: (, Acalamari, Achaemenides, Alansplodge, Aldux, Alex12341000, Amillar, AndrewHowse, Bbb2007, Benjola10, Bluemoose, BrownHairedGirl, Bubba hotep, CambridgeBayWeather, Caomhin, Charlesdrakew, Chzz, Ckatz, Cnyborg, Coolhawks88, Crouch, Swale, Curtis31992, Dahliarose, David Edgar, Deathphoenix, DuncanHill, Edwardx, Eltomzo, Essexman, Gadfium, Galfridus, Garik 11, Geniac, Greebowarrior, Grutness, Gsmgm, Hmains, Inwind, Iosef Ivanov, J Milburn, JForget, JaGa, Jamesslater, Jammydodger, Jeni, Jeremy Bolwell, Jeremythomasflack, Jmlk17, John of Reading, John254, Junglizt1210, Kbthompson, Ken Gallager, Killer kipper, Kingpin13, Ksbrown, Lezek, Loijaa, Lost tourist, LouI, Lupin, MRSC, Magicall95, Malcolmxl5, Marek69, Martarius, Matt.whitby, Mav, MegX, Mendip22, Mikaey, Mindmatrix, Number 57, Olivier, Oneblackline, Orioane, Paedia, Paulleake, Pebkac, Pga23, Piano non troppo, Piperh, Pit-yacker, Pokethesquirrel, Pol098, RFBailey, RandomAct, Regan123, Retew Boy, Rjwilmsi, Rodge500, Rrburke, SchuminWeb, Schzmo, Scribble Monkey, Shar6923, Sheuk, Simply south, Sloman, Smeira, Spymo, Steeev, Stephenb, Tclements1, That Guy, From That Show!, The Anome, Trolleybus fan, Truffles100, Ukbulla, Ulflarsen, W guice, WOSlinker, Warofdreams, Wereon, Wikepping, 245 anonymous edits

Matching_Green *Source*: http://en.wikipedia.org/w/index.php?title=Matching_Green *Contributors*: MRSC, Spymo, Stepheng3, 8 anonymous edits

Matching,_Essex *Source*: http://en.wikipedia.org/w/index.php?title=Matching%2C_Essex *Contributors*: Bencherlite, Crouch, Swale, Gene93k, John of Reading, LilHelpa, MRSC, Spymo, VR6Lee, 2 anonymous edits

Rik_Mayall *Source*: http://en.wikipedia.org/w/index.php?title=Rik_Mayall *Contributors*: 23skidoo, 4twenty42o, 66richardson, A0d1m, AKR619, Acabashi, Al-pocalypse, All Hallow's Wraith, Amicon, Andycjp, Angmering, Anhaidao, Anthony Winward, Antmusic, Arizonasqueeze, Arno Matthias, ArryStreet, Auric, Aurigas, B. Fairbairn, Bart Versieck, Basine, Behind The Wall Of Sleep, Belzediel, Bentley Banana, Bessieb1, BigBouncyBill, BillyH, Bjelleklang, Blurgle Fragle, Bonás, Borobarmy, Bovineboy2008, Bradley0110, CFLeon, Cal T, Cardinal Wurzel, Carrso, Caseywasser, Chaheel Riens, Charlesk, Chris 42, Cinerus, ClicketyBooks, Colonel111, Comedy Dan, CommonsDelinker, Conman2011, Crestville, Curtsurly, Czolgolz, DStoykov, Dailynovet, Danceswithzerglings, Darkieboy236, Darkwarriorblake, DavidFarmbrough, Davidhorman, Deadmanollie, Deb, Dethhura, Didgeman, DionysosProteus, Dobong, Double Blade, DrJos, Dreamspy, DropShadow, Elfroid, Equinoctium, ErinKM, Evlekis, EvocativeIntrigue, FMAFan1990, Feydey, FireRussoFireMantell, FisherQueen, Flaming Ferrari, Flowerpotman, Fourthords, FoxRiver8, Franz-kafka, Freekee, Fys, Fyver528, GHe, Gaius Cornelius, Garion96, Ghostwords, Gkania, Grizzlybear82, Guaka, Gurkha, Gusworld, Halsteadk, Hammersoft, Harryboyles, Helga76, Ian Dunster, Icairns, Interested2, Internetgore, Iridescent, JD554, Jack1956, JamesAM, Jamesgibbon, Jaraalbe, Jdealornodeal, Jeremy Visser, Jim Michael, Jmundo, Joepane, John, Josiah Rowe, K2wiki, Kaleeyed, Karl-Henner, Katherine, Keith D, Kendroche, Kevin Steinhardt, Kevin von always, Kieran1123, Klundarr, Koavf, Kuru, Lanma726, Larroney, Layla12275, LeaveSleaves, Lee M, Leilah, Lenin and McCarthy, Les woodland, Liftarn, LilHelpa, Litefoot, Lllumpy, Longhair, Louiseann76, LuK3, MRSC, Maima, Mar bells87, MarnetteD, Martin S Taylor, MartinSFSA, Mathnawi, Maxim, Mboverload, McAusten, McGeddon, McNoddy, Megastar, MeiChooTan, Michael riber jorgensen, MightyWarrior, MikeJ9919, MikeWattHCP, MilfordBoy991, Mintguy, Mookie89, Mordea, Motrecdsl, Mr. Stabs, Mraakelleher, Music2611, Neural, Nick R, Nidator, Nihiltres, Ningmoss, Nymf, Ocean Shores, Ojophoyimbo, OverlordQ, Owenpeterson, Oxymoron83, Paul A, Paul Benjamin Austin, Peter Chastain, Peter E. James, Phelbasar, Phildogg82, Philip Cross, PhilipC, Piano non troppo, Pigsonthewing, Professor Glass, Propaniac, Pruneau, Rankmole, Ravenclaw, Reaper Eternal, Redman125, Ricky81682, Riddley, Rms125a@hotmail.com, Rogerborg, SM, SausageSandwich, Screenstarx, Sealman, Sean1290, Sendervictorius, Serreel, Severa, Sheridan, Sheriff Bernard, Sheschanged, SidP, Sidecrusher, SietskeEN, Silivrenion, Sims.alec, Sjc, Sjorford, Smith90275, Snigbrook, Solatha, Spearhead, SpeedyGonsales, Spellmaster, Spymo, SteinbDJ, Stephen, Stroppolo, Sugar Bear, TFOWR, TParis, Tangerines, TangoTizerWolfstone, Tarzanboy84, Tassedethe, The Giant Puffin, TheOncomingStorm, Thumperward, Tim!, Tim1988, Timrollpickering, Tinton5, TnaNoZombiesAllowed, TonyW, Tpbradbury, TracyLinkEdnaVelmaPenny, Tregoweth, Treybien, Trident13, Trinak, Unreal7, Valentinian, Van helsing, Vanished user 39948282, Victory93, Vishnava, W4chris, WOSlinker, Wizardman, Woohookitty, WorldWider12, Yrithinnd, Zastrozzi, Zeimusu, 551 anonymous edits

Image Sources, Licenses and Contributors

GNU Free Documentation License Version 1.2, November 2002 Copyright (C) 2000,2001,2002 Free Software Foundation, Inc. 59 Temple Place, Suite 330, Boston, MA 02111-1307 USA Everyone is permitted to copy and distribute verbatim copies of this license document, but changing it is not allowed.

0. PREAMBLE
The purpose of this License is to make a manual, textbook, or other functional and useful document "free" in the sense of freedom: to assure everyone the effective freedom to copy and redistribute it, with or without modifying it, either commercially or noncommercially. Secondarily, this License preserves for the author and publisher a way to get credit for their work, while not being considered responsible for modifications made by others. This License is a kind of "copyleft", which means that derivative works of the document must themselves be free in the same sense. It complements the GNU General Public License, which is a copyleft license designed for free software. We have designed this License in order to use it for manuals for free software, because free software needs free documentation: a free program should come with manuals providing the same freedoms that the software does. But this License is not limited to software manuals; it can be used for any textual work, regardless of subject matter or whether it is published as a printed book. We recommend this License principally for works whose purpose is instruction or reference.

1. APPLICABILITY AND DEFINITIONS
This License applies to any manual or other work, in any medium, that contains a notice placed by the copyright holder saying it can be distributed under the terms of this License. Such a notice grants a world-wide, royalty-free license, unlimited in duration, to use that work under the conditions stated herein. The "Document", below, refers to any such manual or work. Any member of the public is a licensee, and is addressed as "you". You accept the license if you copy, modify or distribute the work in a way requiring permission under copyright law. A "Modified Version" of the Document means any work containing the Document or a portion of it, either copied verbatim, or with modifications and/or translated into another language. A "Secondary Section" is a named appendix or a front-matter section of the Document that deals exclusively with the relationship of the publishers or authors of the Document to the Document's overall subject (or to related matters) and contains nothing that could fall directly within that overall subject. (Thus, if the Document is in part a textbook of mathematics, a Secondary Section may not explain any mathematics.) The relationship could be a matter of historical connection with the subject or with related matters, or of legal, commercial, philosophical, ethical or political position regarding them. The "Invariant Sections" are certain Secondary Sections whose titles are designated, as being those of Invariant Sections, in the notice that says that the Document is released under this License. If a section does not fit the above definition of Secondary then it is not allowed to be designated as Invariant. The Document may contain zero Invariant Sections. If the Document does not identify any Invariant Sections then there are none. The "Cover Texts" are certain short passages of text that are listed, as Front-Cover Texts or Back-Cover Texts, in the notice that says that the Document is released under this License. A Front-Cover Text may be at most 5 words, and a Back-Cover Text may be at most 25 words. A "Transparent" copy of the Document means a machine-readable copy, represented in a format whose specification is available to the general public, that is suitable for revising the document straightforwardly with generic text editors or (for images composed of pixels) generic paint programs or (for drawings) some widely available drawing editor, and that is suitable for input to text formatters or for automatic translation to a variety of formats suitable for input to text formatters. A copy made in an otherwise Transparent file format whose markup, or absence of markup, has been arranged to thwart or discourage subsequent modification by readers is not Transparent. An image format is not Transparent if used for any substantial amount of text. A copy that is not "Transparent" is called "Opaque". Examples of suitable formats for Transparent copies include plain ASCII without markup, Texinfo input format, LaTeX input format, SGML or XML using a publicly available DTD, and standard-conforming simple HTML, PostScript or PDF designed for human modification. Examples of transparent image formats include PNG, XCF and JPG. Opaque formats include proprietary formats that can be read and edited only by proprietary word processors, SGML or XML for which the DTD and/or processing tools are not generally available, and the machine-generated HTML, PostScript or PDF produced by some word processors for output purposes only. "Title Page" means, for a printed book, the title page itself, plus such following pages as are needed to hold, legibly, the material this License requires to appear in the title page. For works in formats which do not have any title page as such, "Title Page" means the text near the most prominent appearance of the work's title, preceding the beginning of the body of the text. A section "Entitled XYZ" means a named subunit of the Document whose title either is precisely XYZ or contains XYZ in parentheses following text that translates XYZ in another language. (Here XYZ stands for a specific section name mentioned below, such as "Acknowledgements", "Dedications", "Endorsements", or "History".) To "Preserve the Title" of such a section when you modify the Document means that it remains a section "Entitled XYZ" according to this definition. The Document may include Warranty Disclaimers next to the notice which states that this License applies to the Document. These Warranty Disclaimers are considered to be included by reference in this License, but only as regards disclaiming warranties: any other implication that these Warranty Disclaimers may have is void and has no effect on the meaning of this License.

2. VERBATIM COPYING
You may copy and distribute the Document in any medium, either commercially or noncommercially, provided that this License, the copyright notices, and the license notice saying this License applies to the Document are reproduced in all copies, and that you add no other conditions whatsoever to those of this License. You may not use technical measures to obstruct or control the reading or further copying of the copies you make or distribute. However, you may accept compensation in exchange for copies. If you distribute a large enough number of copies you must also follow the conditions in section 3. You may also lend copies, under the same conditions stated above, and you may publicly display copies.

3. COPYING IN QUANTITY
If you publish printed copies (or copies in media that commonly have printed covers) of the Document, numbering more than 100, and the Document's license notice requires Cover Texts, you must enclose the copies in covers that carry, clearly and legibly, all these Cover Texts: Front-Cover Texts on the front cover, and Back-Cover Texts on the back cover. Both covers must also clearly and legibly identify you as the publisher of these copies. The front cover must present the full title with all words of the title equally prominent and visible. You may add other material on the covers in addition. Copying with changes limited to the covers, as long as they preserve the title of the Document and satisfy these conditions, can be treated as verbatim copying in other respects. If the required texts for either cover are too voluminous to fit legibly, you should put the first ones listed (as many as fit reasonably) on the actual cover, and continue the rest onto adjacent pages. If you publish or distribute Opaque copies of the Document numbering more than 100, you must either include a machine-readable Transparent copy along with each Opaque copy, or state in or with each Opaque copy a computer-network location from which the general network-using public has access to download using public-standard network protocols a complete Transparent copy of the Document, free of added material. If you use the latter option, you must take reasonably prudent steps, when you begin distribution of Opaque copies in quantity, to ensure that this Transparent copy will remain thus accessible at the stated location until at least one year after the last time you distribute an Opaque copy (directly or through your agents or retailers) of that edition to the public. It is requested, but not required, that you contact the authors of the Document well before redistributing any large number of copies, to give them a chance to provide you with an updated version of the Document.

4. MODIFICATIONS
You may copy and distribute a Modified Version of the Document under the conditions of sections 2 and 3 above, provided that you release the Modified Version under precisely this License, with the Modified Version filling the role of the Document, thus licensing distribution and modification of the Modified Version to whoever possesses a copy of it. In addition, you must do these things in the Modified Version: A. Use in the Title Page (and on the covers, if any) a title distinct from that of the Document, and from those of previous versions (which should, if there were any, be listed in the History section of the Document). You may use the same title as a previous version if the original publisher of that version gives permission. B. List on the Title Page, as authors, one or more persons or entities responsible for authorship of the modifications in the Modified Version, together with at least five of the principal authors of the Document (all of its principal authors, if it has fewer than five), unless they release you from this requirement. C. State on the Title page the name of the publisher of the Modified Version, as the publisher. D. Preserve all the copyright notices of the Document. E. Add an appropriate copyright notice for your modifications adjacent to the other copyright notices. F. Include, immediately after the copyright notices, a license notice giving the public permission to use the Modified Version under the terms of this License, in the form shown in the Addendum below. G. Preserve in that license notice the full lists of Invariant Sections and required Cover Texts given in the Document's license notice. H. Include an unaltered copy of this License. I. Preserve the section Entitled "History", Preserve its Title, and add to it an item stating at least the title, year, new authors, and publisher of the Modified Version as given on the Title Page. If there is no section Entitled "History" in the Document, create one stating the title, year, authors, and publisher of the Document as given on its Title Page, then add an item describing the Modified Version as stated in the previous sentence. J. Preserve the network location, if any, given in the Document for public access to a Transparent copy of the Document, and likewise the network locations given in the Document for previous versions it was based on. These may be placed in the "History" section. You may omit a network location for a work that was published at least four years before the Document itself, or if the original publisher of the version it refers to gives permission. K. For any section Entitled "Acknowledgements" or "Dedications", Preserve the Title of the section, and preserve in the section all the substance and tone of each of the contributor acknowledgements and/or dedications given therein. L. Preserve all the Invariant Sections of the Document, unaltered in their text and in their titles. Section numbers or the equivalent are not considered part of the section titles. M. Delete any section Entitled "Endorsements". Such a section may not be included in the Modified Version. N. Do not retitle any existing section to be Entitled "Endorsements" or to conflict in title with any Invariant Section. O. Preserve any Warranty Disclaimers. If the Modified Version includes new front-matter sections or appendices that qualify as Secondary Sections and contain no material copied from the Document, you may at your option designate some or all of these sections as invariant. To do this, add their titles to the list of Invariant Sections in the Modified Version's license notice. These titles must be distinct from any other section titles. You may add a section Entitled "Endorsements", provided it contains nothing but endorsements of your Modified Version by various parties--for example, statements of peer review or that the text has been approved by an organization as the authoritative definition of a standard. You may add a passage of up to five words as a Front-Cover Text, and a passage of up to 25 words as a Back-Cover Text, to the end of the list of Cover Texts in the Modified Version. Only one passage of Front-Cover Text and one of Back-Cover Text may be added by (or through arrangements made by) any one entity. If the Document already includes a cover text for the same cover, previously added by you or by arrangement made by the same entity you are acting on behalf of, you may not add another; but you may replace the old one, on explicit permission from the previous publisher that added the old one. The author(s) and publisher(s) of the Document do not by this License give permission to use their names for publicity for or to assert or imply endorsement of any Modified Version.

5. COMBINING DOCUMENTS
You may combine the Document with other documents released under this License, under the terms defined in section 4 above for modified versions, provided that you include in the combination all of the Invariant Sections of all of the original documents, unmodified, and list them all as Invariant Sections of your combined work in its license notice, and that you preserve all their Warranty Disclaimers. The combined work need only contain one copy of this License, and multiple identical Invariant Sections may be replaced with a single copy. If there are multiple Invariant Sections with the same name but different contents, make the title of each such section unique by adding at the end of it, in parentheses, the name of the original author or publisher of that section if known, or else a unique number. Make the same adjustment to the section titles in the list of Invariant Sections in the license notice of the combined work. In the combination, you must combine any sections Entitled "History" in the various original documents, forming one section Entitled "History"; likewise combine any sections Entitled "Acknowledgements", and any sections Entitled "Dedications". You must delete all sections Entitled "Endorsements".

6. COLLECTIONS OF DOCUMENTS
You may make a collection consisting of the Document and other documents released under this License, and replace the individual copies of this License in the various documents with a single copy that is included in the collection, provided that you follow the rules of this License for verbatim copying of each of the documents in all other respects. You may extract a single document from such a collection, and distribute it individually under this License, provided you insert a copy of this License into the extracted document, and follow this License in all other respects regarding verbatim copying of that document.

7. AGGREGATION WITH INDEPENDENT WORKS
A compilation of the Document or its derivatives with other separate and independent documents or works, in or on a volume of a storage or distribution medium, is called an "aggregate" if the copyright resulting from the compilation is not used to limit the legal rights of the compilation's users beyond what the individual works permit. When the Document is included in an aggregate, this License does not apply to the other works in the aggregate which are not themselves derivative works of the Document. If the Cover Text requirement of section 3 is applicable to these copies of the Document, then if the Document is less than one half of the entire aggregate, the Document's Cover Texts may be placed on covers that bracket the Document within the aggregate, or the electronic equivalent of covers if the Document is in electronic form. Otherwise they must appear on printed covers that bracket the whole aggregate.

8. TRANSLATION
Translation is considered a kind of modification, so you may distribute translations of the Document under the terms of section 4. Replacing Invariant Sections with translations requires special permission from their copyright holders, but you may include translations of some or all Invariant Sections in addition to the original versions of these Invariant Sections. You may include a translation of this License, and all the license notices in the Document, and any Warranty Disclaimers, provided that you also include the original English version of this License and the original versions of those notices and disclaimers. In case of a disagreement between the translation and the original version of this License or a notice or disclaimer, the original version will prevail. If a section in the Document is Entitled "Acknowledgements", "Dedications", or "History", the requirement (section 4) to Preserve its Title (section 1) will typically require changing the actual title.

9. TERMINATION
You may not copy, modify, sublicense, or distribute the Document except as expressly provided for under this License. Any other attempt to copy, modify, sublicense or distribute the Document is void, and will automatically terminate your rights under this License. However, parties who have received copies, or rights, from you under this License will not have their licenses terminated so long as such parties remain in full compliance.

10. FUTURE REVISIONS OF THIS LICENSE
The Free Software Foundation may publish new, revised versions of the GNU Free Documentation License from time to time. Such new versions will be similar in spirit to the present version, but may differ in detail to address new problems or concerns. See http://www.gnu.org/copyleft/. Each version of the License is given a distinguishing version number. If the Document specifies that a particular numbered version of this License "or any later version" applies to it, you have the option of following the terms and conditions either of that specified version or of any later version that has been published (not as a draft) by the Free Software Foundation. If the Document does not specify a version number of this License, you may choose any version ever published (not as a draft) by the Free Software Foundation. ADDENDUM: How to use this License for your documents To use this License in a document you have written, include a copy of the License in the document and put the following copyright and license notices just after the title page: Copyright (c) YEAR YOUR NAME. Permission is granted to copy, distribute and/or modify this document under the terms of the GNU Free Documentation License, Version 1.2 or any later version published by the Free Software Foundation; with no Invariant Sections, no Front-Cover Texts, and no Back-Cover Texts. A copy of the license is included in the section entitled "GNU Free Documentation License". If you have Invariant Sections, Front-Cover Texts and Back-Cover Texts, replace the "with...Texts." line with this: with the Invariant Sections being LIST THEIR TITLES, with the Front-Cover Texts being LIST, and with the Back-Cover Texts being LIST. If you have Invariant Sections without Cover Texts, or some other combination of the three, merge those two alternatives to suit the situation. If your document contains nontrivial examples of program code, we recommend releasing these examples in parallel under your choice of free software license, such as the GNU General Public License, to permit their use in free software.

Printed by Books on Demand GmbH, Norderstedt / Germany